Unified Field Theory In One Energy Equation

*Dynamic, static, & conversion energy
for electric-magnetic, mass, & waves*

I0474242

Marcus O. Durham, PhD

Realm Research
Tulsa

Unified Field Theory
In One Energy Equation

Contact:
THEWAY Corp.
P.O. Box 33124
Tulsa, OK 74153

www.ThewayCorp.com

Cover Design: Marcus O. Durham, Ph.D.

Printed in United States of America
First printing, Dec 3, 1999
Second printing, November 17, 2011
Revised printing, January 20, 2012

ISBN: 978-1467950701

TO

Rosemary Durham, my spouse, sounding board, and sage, who encourages me in the days and nights of analyzing the tennis ball while developing the concepts and interactions. Her level of consciousness is the highest of anyone personally known.

TABLE OF CONTENTS

Title Page 1

Preface 7

1. What Is a Unified Field Theory? 11
    Background   11
    Simple or not   12
    Energy   13
    Triad Principle   13
    Space Time   14
    Matter   15
    The Math   16
    Unified Field Definition   17
    Energy System   18
    Universal Constants   18
    What Does This Have to Do with You?   19
    What's Next?   20

2. Unified Synopsis 21
    Abstract   21
    Curved Space   21
    Time   23
    Time Space Continuum   23
    Matter   24
    Energy, Force, & Rays   25
    Terms Recap   25
    Unified Field   26
    System Energy   28
    Static Energy   28
    Conversion Energy   29
    Unified Field System Energy   30
    Unified Field System Nodes   31
    Universal Constants   32
    Universal Laws   33
    Construct   33

3. Dynamic Energy in One Curved Space Equation 37
    Abstract   37
    Introduction   37
    Dynamic Energy   38

Curved Space   38
What is the matter?   39
Node   40
Unified Field Dynamic Energy   41
Light   42
Restructure   43
Static   44
Fields   45
Mass Diffusion Field   46
Electric Magnetic Field   47
Wave Constant Field   48
Diffusion   51
Relative Motion   52
Review   53

4. Electromagnetic Fields in One Curved Space Equation 55
Abstract   55
Introduction   55
What is the matter?   56
Curved Space   57
Node   57
Conservation   59
Circuit Analysis and Power   59
Fields   61
Force   62
Field Intensity and Density   62
Diffusion   64
Summary   66
Annex   67

5. Spin on Curved Space Coordinates and Divergence of Time 71
Abstract   71
Introduction   71
Curved Space   72
Manipulation   74
Star Product   75
Fields   76
Time   77
Velocity   79
Speed Limit   79
Closed Path   81
Diffusion   81
Relative Motion   82
Summary   85
Annex   85

6. Static Energy At No Time **87**

    Abstract   87
    Introduction   87
    Static Space   87
    Static Energy   89
    Static Medium Coefficients   90
    Static Capacitance   91
    Static Capacitance for Mass   92
    Static Capacitance for Charge   94
    Static Capacitance for Poles   94
    Summary   96

7. Conversion Energy in Time **97**

    Abstract   97
    Conversion Energy   97
    Electric-Magnetic Resistive Energy   98
    Charge Resistive Energy   98
    Pole Resistive Energy   99
    Diffusion Resistive Energy   100
    Mass Energy Variations   101
    Natural Temperature   103
    A Thing Called Entropy   103
    Summary   104

8. Storage Energy and Second Order Systems **105**

    Abstract   105
    System energy   105
    Storage Charge from Static   106
    Mass: Potential, Viscous, Kinetic   108
    Diffusion: Spring, Damper, Mass   109
    Second Order   110
    Second Order Diffusion   111
    Second Order Electric Charge   113
    Second Order Magnetic Pole   114
    Second Order Physics   115
    Summary   115

9. Universal Constants **117**

    Abstract   117
    Introduction   117
    Triad Constants   118
    Medium Coefficients   119
    Temperature Constant   119
    Other Constants   120

10. Unified Field Summary **121**

Abstract 121
Triad Principle 121
Terms Recap 122
Unified Field Triad Energy 122
Unified Field System Energy 123
Time Space Continuum 124
Curved Space 124
Energy, Force, & Rays 125
That's All 126

A. Temperature and Thermal Transfer.................. **127**
Abstract 127
Introduction 127
Thermal Energy Capacity 127
Thermal Energy 130
Thermal Transfer 131
Transport 132
Summary 134

B. The Calculus, S'il Vous Plaît **135**
Abstract 135
Introduction 135
Spectrum and Sampling 136
Spectrum Radiation 138

References.. **141**

About the Author.. **145**
Marcus O. Durham 145

⇐ ⇑ ⇒

PREFACE

Everything we know is developed from something we have read, heard, or seen. Therefore, these other thoughts necessarily influence what we write. To the best of our knowledge, we have given specific credit where appropriate.

Rather than footnotes or references, we have listed the works that have provided significant information in one way or another, since this is often in concepts rather than specific relationships. Papers resulting from the theory are also noted. The concepts were also used in the books written for university classes taught.

Statements that are attributed to us are things we have used commonly and do not recall seeing from someone else. Others obviously may have similar thoughts. If we have made an oversight in any credits, we apologize and we would appreciate your comments.

Because of the extensive equations with myriad subscripts, some scripts may have been improperly noted. That should not influence the understanding of the concepts.

Exceptional effort has been made to provide an accurate, true representation of the concepts. Numerous scientists and proofreaders have evaluated the results. If some deviation is found, please contact the publisher, so future editions will have the benefit of further review.

1

WHAT IS A UNIFIED FIELD THEORY?

Thought
*The unified field incorporates all
of matter, space, and time.*
MOD

Background _____

What is energy? What are the components of energy? What are the components of matter? What are the components of space? What are the components of time? Can all these items be combined in one comprehensive relationship?

A unified field is a comprehensive definition of how all matter, space, and time fits together. What is that interaction?

There has been considerable interest for centuries in developing a simple, but comprehensive relationship for natural, physical systems including electric and magnetic, mass, and wave interactions, even before these concepts were well defined.

Energy is the forum for these issues. Light is the ultimate manifestation of energy. Light has been recognized since the dawning of man. There had been a progression of understanding through history, but none of these ideas were comprehensive.

Tangible identification of light is first documented in modern literature during the 1600's when Robert Hooke proposed that light is a wave. Christian Huygens refined the concept to include individual wavelets. Sir Isaac Newton contended that light radiated particles which were called corpuscular. In the late 1800's James Clerk Maxwell proposed that light was a series of electric and magnetic interactions. Then Max Planck and Albert Einstein proposed that light was a quanta, or individual particles that were emitted.

Each and every one of these concepts is readily identifiable from a unified field energy equation.

Simple or not _____

Einstein gave a simple, powerful, comprehensive relationship for mass when he proposed its relationship to energy. A similar, single correlation has not been shown for the electrical and magnetic spectrum, until these principles were proposed last century by Durham.

Maxwell gave a group of field correlations for the electric and magnetic realm. This has traditionally been the foundation for developing virtually all electrical and magnetic arguments. Although elegant in form, the mathematics is considerably more complex than the very familiar Einstein relationship that is commonly known to the general public, but perhaps not well understood.

To be simple, the mathematics must allow the relationship to be used for absolute, differential, or relativistic perspectives. Absolute perspective is conventional calculations that commonly use definite numbers such as a checking account. Differential perspective looks as the differences or changes in values during some time frame. Relativistic perspective is the sophisticated system of comparing motion to another activity.

To be comprehensive, the relationship must encompass both nodes or lumped points and fields or distributed elements in a fluid. Nodes are a very defined location or connection where all effort is concentrated. The effort is lumped at the point. In many other circumstances the effort is scattered or dispersed. This distribution results in a field which is dispersed through a fluid consisting of a liquid or gas such as air.

These criteria are contained in a unified field energy relationship.

Energy _____

Energy is the common basis for conversion of all systems. But what is energy? It is a measure of work, heat, and loss. Energy is a force that moves through a distance. Energy can change forms, but always remains; it is neither created nor destroyed.

The foundation is described by the well-known Conservation of Energy.

> *Energy cannot be created or destroyed, but may only change form.*

Alternately, it is stated as the sum of the energy in a closed system, without outside influence, is zero.

$$\Sigma E = 0$$

Some of the derivatives of the Conservation of Energy include the Conservation of Momentum, Conservation of Mass, Conservation of Charge, and Conservation of Magnetics. Each of these conservation states is simply a special case that is derived when other terms are held constant in the energy relationship.

All these concepts must be directly obtained from a unified field energy equation.

Triad Principle _____

Natural physical systems yield a principle for research models. The fundamental principle has been uniformly observed and ascertained for all physical relationships studied. This statement was published in numerous peer reviewed papers and in books over the past twenty-five years.

> *Triad Principle – Any item that can be uniquely identified can be further explained with three components.*

> *Corollary – Two of the items will appear similar and the third will appear orthogonal.*

A second order system is used to describe natural phenomenon. A second order system is one that has three terms, one of which has a second time component. The second order relationship is an obvious representation of the triad principle.

The necessary terms for any natural energy system can be identified using this grouping of three quantities. If a discussion of a system has either more or fewer items, it is either a combination of unique terms, or an inadequately explained or inadequately defined system.

The Triad Principle, therefore, provides a mechanism for research into areas that are unknown.

These triad criteria are definitive of a unified field relationship.

Space Time_____

Energy is the concept for conversion between physical systems. Energy or light consists of three components - space, time, and matter. It really is that simple. According to the Triad Principle, each of these can progressively be broken down into three aspects or states.

The first of the energy components, space, has three measures – motion, ray, and volume. Motion is a vector that tells the distance and direction an item is moving. Ray is the length of the arm acting on the force. Volume is the physical dimension measure.

Volume has three dimensions. In a flat-earth world, the three dimensions are straight lines that are perpendicular. In reality the earth and all within its realm is a spherical type space. Therefore, the three dimensions must be described for a curved space. A coordinate system that uses a system more conducive to airplane and space travel is introduced. The same system works at the sub-atomic level.

The second energy component is time. The space time relationship obviously has a way to measure events and the duration between the events. Time has three manifestations. In static situations, there is no movement, so time has no variation. Constant time is called aeon and is represented by the number 1. The time that measures most activity is cyclic and is called chronos. Cyclic or motion time is represented by 't_t'. The third time manifestation is directly related to how mass reacts to space. Space time is reference or seasonal and is called kairos 't_r'.

The product of the three measures of space called motion, ray, and volume divided by space direction and the reference, seasonal time is called diffusion. Diffusion is simply the explanation of the space and time interaction.

Space and time are inseparable. Together they form a continuum.

Matter _____

The third component of energy is matter. Matter is the fabric of which physical items are made. Matter is three regents – mass 'm', charge 'q', and magnetism 'p'. Each regent has its own energy domain.

Mass is the thing that can be perceived with the five senses of see, hear, touch, taste, and smell. Mass is uniquely coupled to the space time concept. The space time interaction is called diffusion 'D'. Mass operates in the mechanical and gravitational domain.

Charge is the measure of electrical activity. Charge, as electrons and protons, is a key ingredient of chemical and nuclear phenomenon. Charge is the electrical energy domain.

Magnetism is the measure of magnetic attraction. Magnetism is observed from a pair of poles. Magnetism is inextricably linked to charge when either is in motion. Magnetism is the magnetic energy domain. Together charge and magnetism make the electro-magnetic spectrum.

In the fundamental perception, each of these regents occurs at a point or node. All the energy is lumped at one location. In many instances the energy is dispersed in a fluid made up of liquid or gas. In fields, the regent effects are distributed over a volume.

Matter, space, time moves in waves 'w', oscillations, or vibration motion. There is energy from the oscillation when coupled to a constant of the universe 'h_p'.

This perception of natural physical systems provides a principle for a unified research model.

The Math _____

To this point all the terms have been introduced.

1. Energy consists of space, time, and matter.
2. Space consists of three dimensions of length, width, and height.
3. Dynamic volume consists of motion, lever and space reference distance vectors.
4. Time consists of constant, cyclic motion, and seasonal reference.
5. Matter consists of mass, electric charge, and magnetism.

Next, the concepts are combined to form a unified field. That step necessarily involves mathematics. The model is explained with only multiplication and division applied to the Conservation of Energy.

Mathematics is nothing more than symbols to illustrate the relationship between values. Virtually everyone is familiar with the symbols for addition, multiplication, and division. The symbols link numbers, the calculation can be completed. Often the numbers are also replaced by symbols to show that different numbers can be used in the calculation. That is as complex as the mathematics is in the unified field.

All calculations discussed are changes in values of variables from a reference. The relations represent the instantaneous and peak values. Therefore, calculus notation is superfluous. Derived concepts and average values may depend on the calculus, but these will not be developed in this treatise.

Unified Field Definition _____

The desire for a unified field relationship is to tie together the concepts of matter, space, and time.

This elusive goal is realized with a very straightforward concept. As would be expected, the notion is expressed with three terms.

1. The first term is the multiplication of mass and space time diffusion divided by cyclic time.

$$E = \frac{m\,D}{t_t}$$

2. The second term is the product of magnetism and charge divided by cyclic time.

$$E = \frac{p\,q}{t_t}$$

3. The third term is the product of a universal constant and the number of waves divided by cyclic time.

$$E = \frac{h_p w}{t_t}$$

Following the conservation of energy, these three terms are added together to yield the Unified Field. The three terms of the Unified Field are in triad form and are also called the dynamic energy.

$$E = \frac{m\,D}{t_t} + \frac{p\,q}{t_t} + \frac{h_p w}{t_t}$$

The fundamental dynamic energy was accomplished with only seven parameters. Think about it. The reckoning that describes all of matter has only seven items that are combined.

The simplicity of the relationship expresses all the elusive terms for light that have been proposed throughout history. This is the handle for how the universe is wired. Whether looking in space, at the book on your desk, or at sub-atomic nuclear particles, all operates according to this simple, elegant summation of matter.

The fundamental concepts help to see the order of the natural, physical world. The application of the principles is for the scientists, mathematicians, and engineers.

Energy System_____

The discussion was the dynamic energy, which describes the interaction of matter. System energy includes the states of the energy under different time conditions.

An energy system has three states - dynamic, static, and conversion. Dynamic energy as noted involves cyclic time. Static energy develops without time. Conversion energy is involved when one of the regent domains is changed to a different domain. As would be expected, the static energy has three terms and the conversion energy has three terms.

The System concept of the Unified Field has the dynamic or Triad energy equal to the sum of the static and conversion energy terms.

Universal Constants _____

In nature, there exists very few absolute, unchanging numbers or values. These are called universal constants.

Three constants are linked to the Triad energy. These are the pi – the number associated with a circle or cycle, the speed of light, and the energy associated with the number of waves or cycles.

Three other constants are linked to the static energy. Another constant is associated with the conversion energy and temperature.

There are simply seven universal constants necessary to completely define the fixed values of nature.

What Does This Have to Do with You?

So what does this science discussion have to do with you? Since the unified field shows how systems interact, it is possible to reconfigure understanding of complex systems and processes.

Consider health and medical applications. The relationships explain the interaction of molecules, cells, and organisms.

Consider energy systems. Alternative types and mechanisms can be configured to be more creative, efficient, and smaller.

Consider philosophy, psychology, and entertainment. Knowing what stimulates the human condition radically changes how we interact.

Consider travel. The speed and type of transportation can be increased.

Consider communications. The present data overload can be handled using techniques that filter unnecessary information and provide only positive, desirable data.

The realization of the interaction of matter, space, and time provides a vehicle for positive feedback in virtually every area of science endeavor.

What's Next?_____

The well-informed. The chapter provides the fundamentals so that a reasonably well-informed individual can comprehend the elegance of the unified field. This chapter has introduced the unified field as a language concept with limited use of mathematics. The chapter is conceptual so that understanding is not restricted by equations and elegant symbolism. The unified field explains how all matter, space and time fits together.

Scientist understanding. The next chapter gives the general information so that a scientist or mathematician has a good understanding of the field. The chapter will provide the foundations, terms, and equations of the unified field with very limited discussion about the interactions. This is a complete overview with essential terms and models. Comprehension of this chapter gives the basis for broad understanding and application of the concepts.

In depth. The remaining chapters are intended for the scientist that desires to put it all together. The details illustrate how the unified field relates to energy and force in various systems. These chapters show how conventional Newtonian and relativistic Einsteinian concepts can be obtained from the Unified Field.

$$\Leftarrow \Uparrow \Rightarrow$$

2

UNIFIED SYNOPSIS

Thought
*Unified field incorporates all
of matter, space, and time.*
MOD

Abstract _____

The chapter provides an overview of the unified system for the scientist. The concepts introduced are curved space, dynamic volume gradient and the space time continuum. Dynamic energy is the model for the Unified Field Triad energy. Static energy and conversion energy are added to create the System model. The universal constants are also identified.

Curved Space _____

Physical systems operate in a curved space. This has traditionally created difficulty because most analysis uses a rectangular Cartesian coordinate system. The complexity of Maxwell's suite arises from the calculus on a rectangular system.

Similarly, Einstein's relativity has conceptual challenges because of the coordinate system and the number of terms necessary to describe an adjacent, relative position using tensors.

An alternative coordinate system greatly improves conceptual illustration of these physical networks. A curved coordinate system resolves both gravitational and electromagnetic problems.

The effects of time on curved space create a divergence that redefines position regent. By judicious application of these principles, the unified driving energy for natural, physical systems can be resolved. A method of mathematics is developed.

Space and motion in our perspective is a three-dimensional spheroid with measurements on the surface. The three types of curved space distance or dimensions are motion, torque lever or rotational, and space reference.

Motion distance or wavelength 'd_t' is measured as a latitude line tangential around a sphere. The lever arm is a ray 'b_{rs}' that is projected onto the reference or relative axis 'b_r' and the longitude 'b_s'. The motion distance passes through the end point of the ray. The ray is not along an axis and the length will vary with the contour.

Space distance 's' has three components that encompass a distance along a latitude or tangential 's_t', a distance up a longitude or standing coordinate 's_s', and a distance along a reference 's_r'. The reference may be oriented along an internal Cartesian coordinate 'x, y, z' to provide a transform into that reference system.

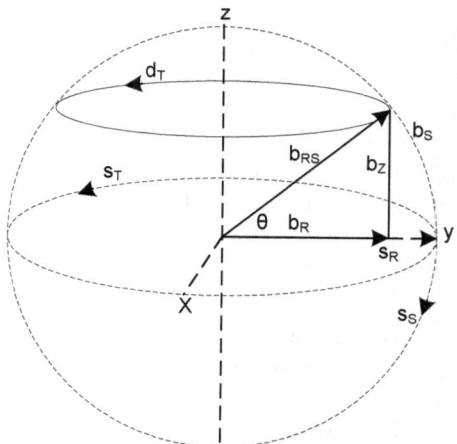

Figure 1 – Curved Space Vectors

The volume does not have to be uniform. The volume can be represented as the average area of the surface of a Riemann sphere and the dot product of the volume distance. In a uniform circular space, the curved space area and volume would have the following average values.

$$A_r = s_s \times s_t$$

$$A_r = 4\pi s_r^2$$

Where,

$$V_r = s_s \times s_t \cdot s_r$$

$$V_r = 4\pi s_r^2 \cdot s_r$$

$$V_r = \frac{4}{3}\pi s_r^3$$

Time _____

Time has three manifestations. Constant time is represented by the factor 1. Motion time is cyclic and is represented by subscript t 't_t'. Reference time is long term seasonal and is represented by subscript r 't_r'. Further, the subscript illustrates the axis of the action.

Time is not isolated in context. Time is directly coupled to space. Space and time are inseparable. Together they form a continuum.

Time Space Continuum_____

The dynamic volume is defined by the ray, motion, and space reference distance vectors.

$$V_r = b_{rs} \times d_t \cdot s_r$$

A field is simply energy that is distributed in a fluid, whether liquid or gas. Interestingly, a recurrent concept is the dynamic volume gradient 'dvg'.

$$dvg = \frac{b_{rs} \times d_t \cdot s_r}{s_r}$$

The notion of the dynamic volume gradient over reference time 't_r' is now defined as diffusion 'D'.

$$D = \frac{b_{rs} \times d_t \cdot s_r}{t_r \, s_r}$$

The space reference is very intentionally maintained in both the numerator and denominator. It is necessary in order to have volume in the numerator and it is necessary in the denominator to identify fluids and fields problems. In some issues, the gradient factor appears to cancel, which would dimensionally yield an area rather than volume.

The rate of diffusion over motion time 't_t' becomes the product of reference velocity and motion velocity.

$$\frac{D}{t_t} = \frac{b_{rs} \times d_t \cdot s_r}{t_r \, t_t \, s_r} = u_{rt} u_t$$

In Newtonian motion, the diffusion rate is velocity squared. However, at the boundary interface it is the speed of light squared, a critical component of Einsteinian relative motion.

$$\frac{D}{t_t}\bigg|_{lim} = c^2$$

Matter_____

The third component of energy is matter. Matter is the fabric of which physical items are made. Matter is three regents – mass 'm_x', charge 'q_y', and magnetism 'p_z'. Each regent has its own energy domain.

Mass is the thing that can be perceived with the five senses. Mass is uniquely coupled to the space time concept. The space time interaction is called diffusion 'D'. Mass operates in the mechanical and gravitational domain.

Charge is the measure of activity in the electrical domain. Magnetism is the measure of magnetic attraction. Magnetism is inextricably linked to charge when either is in motion. Together charge and magnetism make the electromagnetic spectrum.

Matter, space, and time moves in waves 'w', oscillations, or vibration motion. There is energy from the oscillation when coupled to Planck's constant of the universe 'h_p'.

Energy, Force, & Rays _____

One concept of energy is the product of force through a ray distance. When that necessarily is expanded to incorporate spherical space vectors, then linear and rotational energy is defined by the star product.

$$
\begin{aligned}
E &= b_{rs} * F_r \\
&= b_r \cdot F_r + (b_s \times F_r)_t \\
&= b_r \cdot F_r + \left(\theta_{rs} b_j \times F_r\right)_t \\
&= b_r \cdot F_r + (T_t \theta_{rs})_t
\end{aligned}
$$

Each of the energy field terms has a ray distance vector 'b_{rs}' component. The remaining parameters of the term make up the force. The result is eight defined forces and the conversion temperature.

Terms Recap_____

To this point all the terms have been introduced.

1. Energy consists of space, time, and matter.
2. Spheroidal space consists of three dimensions of tangential 's_t', standing 's_s', and radial reference, 's_r'.

3. Dynamic volume incorporates three unique space distance types of lever or ray 'b_{rs}', motive or wavelength 'd_t', and volume space 's_r'.

4. Similarly, time has three manifestations of constant '$t=1$', cyclic motion 't_t', and seasonal reference 't_r'.

5. Then, matter has three regents of mass 'm', electric charge 'q', and magnetic pole strength 'p'.

6. The life energy is the number of waves 'w' and Planck's constant of the universe 'h_p'.

Next, the concepts are combined to form a unified field.

Unified Field _____

The desire for a unified field relationship is to tie together the concepts of matter, space, and time.

This elusive goal is realized with a very straightforward concept. As would be expected, the notion is expressed with three terms.

1. The first term, Law of Mass-Diffusion energy, is the multiplication of mass and space time diffusion divided by cyclic time.

$$E(d) = \frac{m_x D}{t_t}$$

2. The second term, Law of Electric-Magnetic energy, is the product of magnetism and charge divided by cyclic time.

$$E(d) = \frac{p_z q_y}{t_t}$$

3. The third term, Law of Wave-Constant, energy is the product of Planck's universal constant and the number of waves divided by cyclic time.

$$E(d) = \frac{h_p w}{t_t}$$

Following the conservation of energy, these three terms are summed to yield the Unified Field energy. The three terms of the Unified Field are in triad form operating at a node.

Unified Field Triad Energy, node model

$$E(U) = \frac{m_x D}{t_t} + \frac{p_z q_y}{t_t} + \frac{h_p w}{t_t}$$

The fundamental Unified Field Triad energy was accomplished with only seven parameters to describe the operation of the universe. Think about it. The reckoning that describes all of matter has only seven items that are combined.

In the fundamental perception, each of the regents occurs at a point or node. All the energy is lumped at one location. In many instances, the energy is dispersed in a fluid made up of liquid or gas. In fields, the regent effects are distributed over a volume.

Triad energy in node form describes the activity of regents 'm, p, q' over motion time 't_t'. Triad energy in field and fluid form expands to incorporate the dynamic volume gradient.

Unified Field Triad Energy, field model

$$E(U) = \frac{m_x}{t_t} \frac{b_{rs} d_t s_r}{t_r s_r} + \frac{p_z q_y}{t_t} \frac{b_{rs} d_t s_r}{s_s s_t s_r} + \frac{h_p w}{t_t} \frac{b_{rs} d_t s_r}{s_s s_t s_r}$$

The Unified Field Triad energy incorporates all of matter, space, and time. At the boundary interface, it provides the definition of light.

System Energy _____

The Unified Field energy is the dynamic energy of a system. Dynamic energy describes the interaction of matter.

An energy system has three states - dynamic, static, and conversion. Dynamic energy as noted involves cyclic time. Static energy develops without time. Conversion energy is involved when one of the regent domains is changed to a different domain. As would be expected, the static energy has three terms and the conversion energy has three terms.

The System concept of the Unified Field has the dynamic or Triad energy equal to the sum of the static and conversion energy terms.

Static Energy _____

Static energy is the product between two regents of similar kind, the medium coefficient, the ray vector, and the space reference vector over the space volume.

$$E(k) = \frac{b_{xs}k_m m_0\, m_x\, s_x}{s_s s_t s_x} + \frac{b_{ys}k_q q_0\, q_y\, s_y}{s_s s_t s_y} + \frac{b_{zs}k_p p_0\, p_z\, s_z}{s_s s_t s_z}$$

The node or lumped parameter form has the space and medium coefficient grouped together.

$$E(k) = \frac{m_0\, m_x}{C_m} + \frac{q_0\, q_y}{C_q} + \frac{p_0\, p_z}{L_p}$$

The coefficients are three of the universal constants.

Coefficient	k_0	Value	Std units	Alt units
k_m	$4\pi \times \gamma_0$	$4\pi \times 6.67384 \times 10^{-11}$	N-m^2/kg^2	(m^2/s)-m/kg-s
k_q	$1/\varepsilon_0$	$36\pi \times 10^9$	N-m^2/cb^2	wb-m/cb-s
k_p	$1/\mu_0$	$1/(4\pi \times 10^{-7})$	N-m^2/wb^2	cb-m/wb-s

Conversion Energy _____

Conservation of energy declares that energy is neither created nor destroyed, but may only change form. The energy of each regent can be converted to another form. Electric-magnetic conversion comes from static energy, while mass-diffusion conversion comes from dynamic energy.

The conversion relationship of the electromagnetic term is defined by the static type energy 'q' expanded over time. The lumped parameter or node relationship combines all terms except the charges over time.

$$E(r) = \frac{b_{ys} r_q q_0\, q_y\, s_y}{t_t s_s s_t s_y}$$

$$E(r) = R_q \frac{q_0\, q_y}{t_t}$$

A later chapter illustrates the magnetic equivalent of the charge conversion. It is simply a rearrangement of the magnetism in lieu of charge.

The conversion of energy from forms other than electric-magnetic is contained in the mass-diffusion 'E(r)' relationship.

$$E(r) = \frac{m_x}{t_t} \frac{b_{rs} d_t s_x}{t_r s_x}$$

$$E(r) = B \frac{b_{rs} d_t}{t_t}$$

Where,

$$B = \frac{m_x s_x}{t_r s_x}$$

The final term in the universal energy equation is the natural temperature 'T', Boltzmann's constant 'h_B', and the number of interactions 'N'. At the molecular level, N is the number of molecules.

$$E(T) = N\, h_B\, T$$

There is opposition or resistance 'R' or 'B' to the conversion by the system giving up the energy 'E(r)'. The conversion of energy to another form results in work 'W' or accomplishment which is recovered. However, each conversion has a loss that is payback to the universe. The energy loss is converted to entropy 'S' and temperature 'T'.

$$E(r) - S\,T = W$$

Without entropy, the system is ideal. Entropy is the measure of non-productive work. The entropy relationship is not part of the fundamental unified equation. However, it is necessary to yield the work that is available to another system. Therefore, it will be shown parenthetical.

The conversion energy available is the sum of the resistance energy plus the natural temperature minus loss.

$$E(r) = \frac{b_{ys} r_q q_0\, q_y\, s_y}{t_t s_s s_t s_y} + \frac{m_x s_x}{t_r s_x}\frac{b_{rs} d_t}{t_t} + N\, h_B\, T\, (-ST)$$

Unified Field System Energy _____

The Unified Field System (UFS) energy incorporates the states of energy to explain activity in a system. The system energy states are dynamic, static, and conversion. System energy is the three dynamic energy terms (d), the three static energy terms (k), and the three conversion terms (c).

$$dynamic\ energy = static\ energy + conversion\ energy$$

The Unified Field System energy is one equation of three regent terms for three states of energy.

Unified Field System Energy, field model

$$\frac{m_x}{t_t}\frac{b_{rs}d_t s_r}{t_r s_r} + \frac{p_z q_y}{t_t}\frac{b_{rs}d_t s_r}{s_s s_t s_r} + \frac{h_p w}{t_t}\frac{b_{rs}d_t s_r}{s_s s_t s_r} =$$

$$\frac{b_{xs}k_m m_0\, m_x\, s_x}{s_s s_t s_x} + \frac{b_{ys}k_q q_0\, q_y\, s_y}{s_s s_t s_y} + \frac{b_{zs}k_p p_0\, p_z\, s_z}{s_s s_t s_z} +$$

$$\frac{b_{ys}r_q q_0\, q_y\, s_y}{t_t s_s s_t s_y} + \frac{m_x s_x}{t_r s_x}\frac{b_{xs}d_t}{t_t} + N\, h_B\, T\, (-S\,T)$$

As a unified field, these energy relationships are valid for any and every analytical perspective, whether cosmic or sub-atomic and Newtonian or Einsteinian. Units and other restrictive factors have been avoided to preclude limitation of perception.

The unified field is expressed in the most fundamental terms to allow arrangement for any system whether electrical, mechanical, or chemical.

Unified Field System Nodes _____

The field equation is defined by every variable or constant relevant to the energy terms.

In many calculations the parameters are lumped together to describe a particular problem or system. One implementation of the Unified Field System equation in node form is shown.

Unified Field System Energy, node model

$$\frac{p_z q_y}{t_t} + \frac{m_x D}{t_t} + \frac{h_p w}{t_t} =$$

$$\frac{m_0\, m_x}{C_m} + \frac{q_0\, q_y}{C_q} + \frac{p_0\, p_z}{L_p} +$$

$$R_q \frac{q_0\, q_y}{t_t} + B \frac{b_{rs} d_t}{t_t} + N\, h_B\, T\, (-S\, T)$$

Other implementations can exist, dependent on the system being developed and how the dynamic volume gradient terms are combined.

Universal Constants _____

The universal triad constants are imbedded in the Unified Field Triad energy. As would be expected, there are three and only three constants in the fundamental energy.

Pi 'π' is the ratio of the circumference to the diameter for a circle.

Speed of light 'c' is a constant circle constraint that forms a spheroidal boundary interface on velocity. So, it is the limiting velocity at the interface boundary. The speed of light binds cyclic time to motive distance, called wavelength.

Planck's constant 'h_p' is the energy value of a wave or quanta.

The complete suite of universal system constants arise from implementation of the Unified Field System equation. Only seven fundamental constants exist. These comprise the three triad constants, three medium constants, and the temperature constant.

Permittivity 'ε' is the medium parameter for electric charge.

Permeability 'μ' is the medium parameter for magnetic poles.

Universal gravitation 'γ' is the medium parameter for mass.

Boltzmann's constant 'h_B' is the energy value of temperature.

Universal Laws_____

There are seven energy laws, two mathematical circuital laws and one diffusion law associated with the unified field.

Law of electric-magnetic energy

Law of mass-diffusion energy

Law of wave-constant energy

Law of static energy

Law of conversion energy

Law of temperature-constant energy

Law of conservation of energy

Law of circuital operator

Law of circuital energy

Law of diffusion

Construct _____

The well-informed. The first chapter provided the fundamentals so that a reasonably well-informed individual can comprehend the elegance of the unified field. The chapter introduced the unified field as a language concept with limited use of mathematics. The chapter is conceptual so that understanding is not restricted by equations and elegant symbolism. The unified field explains how all matter, space and time fits together.

Scientist understanding. This chapter gives the general information so that a scientist, mathematician, or engineer has a good understanding of the field. The chapter provides the foundations, terms, and equations of the unified field with very limited discussion about the interactions. This is a complete overview with essential terms and models. Comprehension of this chapter gives the basis for broad understanding and application of the concepts.

In depth. The remaining chapters are intended for the scientist that desires to put it all together. The details illustrate how the unified field relates to energy and force in various systems. These chapters show how conventional Newtonian and relativistic Einsteinian concepts can be obtained from the Unified Field.

All concepts are explained and illustrated with more detail in subsequent chapters. Each chapter is stand alone. All the concepts for that aspect are included together. Consequently, there is a limited amount of restatement of some fundamental ideas. Nevertheless, this does aid in application of the specifics for the chapter.

Three chapters are used to illustrate the Unified Field Triad energy.

The initial chapter gets to the heart of the matter using dynamic energy to describe a Unified Field Triad equation. The chapter looks at mass, electric-magnetic, and wave terms. Einstein's as well as Newton's physics are contained in the mass term.

The next chapter takes the electric and magnetic term and expands it to encompass the entire study of electrical and magnetic phenomenon. From this one term, Maxwell's four laws are derived. In reality, the numerous laws of electromagnetic analysis are all based on the one unified Law of Electric-Magnetic energy.

The subsequent chapter develops an elegantly simple spherical coordinate system that explains curved space without the necessity of the calculus. From the coordinate system, curved space and divergence of time are readily observable. The spherical system is comprehensive for cosmic, global, and subatomic analysis. Euler's relationships are a subset of the curved space, spherical coordinate system.

System energy is described in the next three chapters. System energy includes static, conversion, and storage energy.

Static energy for mass, charge, and magnetism, is developed. From this concept, potential energy for mass as well as electric and magnetic energy is illustrated.

Conversion between electrical and magnetic to mechanical energy is identified in the following chapter. Conversion involving mass systems is embedded in the dynamic energy mass term. The energy of natural temperature is the final term. The conversion minus the entropy effect yields work.

Storage energy is developed from other energy terms and is not unique. However, it is required to describe system activity. The form of system energy is a second order. The solution of the second order is elucidated. Every known energy system is second order in nature at its root. Therefore, a standard solution is presented.

The universal constants of nature are identified in another chapter. Only three constants are unique to the Unified Field Triad while seven constants are unique to the Unified Field System.

The final chapter is a summary of the Unified Field energy.

Two annexes are supplied to further explain temperature and spectrum distribution.

Temperature is a common manifestation of energy. Temperature and energy transport are discussed in Annex A. This discussion is not fundamental to the unified field.

Spectrum distribution is commonly used rather than the discrete analysis used in this treatise. Spectrum influence is discussed in

Annex B. Spectrum analysis arises from uncertainty. Uncertainty arises from incomplete description of the problem. The Unified Field mediates much of the need for spectrum analysis and uncertainty.

3

DYNAMIC ENERGY IN ONE CURVED SPACE EQUATION

> Thought
> *Dynamic energy is regent activity*
> *with cyclic time in the denominator.*

Abstract _____

Physical systems can be correlated through energy. The treatise proposes several relationships to define dynamic energy. These include manifestations of mass-diffusion, pole-charge, and wave-constant over time. The fundamental building blocks are described. Then the surface dynamic energy laws are related into one unified field equation. By judicious application of these principles, the driving energy for natural physical problems can be resolved. A single relationship correlates mass to energy, Maxwell's equations, and the vibrations of string theory.

In this chapter, dynamic energy is expanded to include mass and waves along with charge and magnetism. The next chapter combines electric and magnetic energy into one equation. The subsequent chapter of the trilogy describes spheroidal space coordinates and divergence of time. The principles have extended correlation to other laws such as static and thermal energy.

Introduction _____

There has long been an effort to correlate various energy and forces in physical systems. Each time a new relationship is developed, more questions arise. Nevertheless, system theory requires a determination of interactions. In order to relate conversions between disciplines, this chapter combines the diverse relationships into one brief procedure.

From the principle, numerous concepts can be explained. Energy is used as the conversion mechanism. Energy can be described in terms of the fundamental matter, time, and space. First surface dynamic energy is defined. Then space is added to the system to describe fields. Applications contain strategic examples of combinations of energy relationships.

Dynamic Energy _____

Dynamic energy is regent activity with cyclic time in the denominator. Dynamic energy exists for mass, electro-magnetics, and waves. The relationship for these three manifestations is developed from the regents of matter. The activity is on the surface of curved space and can be described with the spheroidal coordinate system.

Energy relationships are changing values related to a reference.

Time is a denominator function creating a rate. The resulting energy is an instantaneous peak. Therefore, calculus representation is superfluous. However, derived concepts and average values may depend on the calculus, but these will not bew developed.

Curved Space_____

Curved space analysis using a spheroidal coordinate system is developed in a later chapter. The figure is included here for reference to the subscripts.

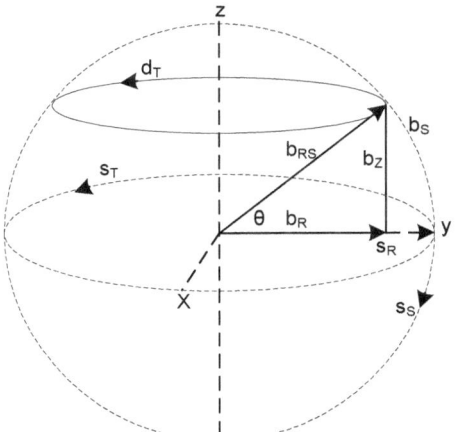

Figure 1 – Curved Space Vectors

Notice that the internal axes are a projection of the surface axes. For example, 's_y' is a projection of 'd_t', and 's_z' is a projection of 's_s'. The general reference subscript 'r' represents the 'x' or 'y' internal axis, depending on the system.

Each vector is noted by a subscript that corresponds to a surface axis or internal axis. All variables are vectors, since they operate in a direction along an axis and have magnitude.

What is the matter? _____

Matter consists of three regents – mass 'm', charge 'q', and magnetism 'p'. In the fundamental perception, each of these occurs at a point or node. In fields, their effects are distributed over a volume. The magnetism is then referred to as magnetic flux, 'Φ'.

Each regent has its own energy domain. Mass yields mechanical or gravitational energy, charge yields electric energy, and magnetic poles yield magnetic energy.

Three values are implemented when converting the regents to energy. A constant 'h_P' is used for unit conversion. A discrete integer 'w' is used to count the number of waves, cycles, spins, or vibrations. A variable 'D' is the diffusion, or rate of dynamic volume gradient to a concentration.

The dynamic energy equation is an energy conversion manifest by combining the matter regents over cyclic time, 't_t'. This yields a rate of change for the regents and the variable. Three unique terms can completely encompass all system concepts. These are pole-charge, mass-diffusion, and wave-constant.

Node

The first development will be for the regents at a point or node. Later, the field model is developed for diffusion in space.

The first dynamic energy manifestation is mass-motion conversion.

Law of mass-diffusion energy, node model
Mass-diffusion energy is the product of the mass and the diffusion over time.

$$E(d) = \frac{m_x D}{t_t}$$

This product involves one regent of matter. Diffusion, as detailed later is the change of dynamic volume gradient over reference, seasonal time. Diffusion is the space-time interaction term.

The second dynamic energy manifestation is further developed in the next chapter as the law of electric-magnetic energy.

Law of electric-magnetic energy, node model
Electric-magnetic energy is the product of magnetism and charge over time.

$$E(d) = \frac{p_z q_y}{t_t}$$

This product is two regents of matter over time.

The tertiary dynamic energy manifestation, which includes the concept of harmonics and bundles, is the wave energy conversion.

Law of wave-constant energy, node model
Wave energy is the product of Planck's universal constant and the number of waves (cycles) over time.

$$E(d) = \frac{h_p w}{t_t}$$

This is the product of a constant and a discrete number over time. The number of waves is not continuously differential as an analog value. It is a difference of discrete values.

Unified Field Dynamic Energy _____

The three manifestations encompass the diverse components of dynamic energy combined in one expression of seven terms. The expression is the unified field equation, node model.

Equation of dynamic energy, node model
Dynamic energy is the sum of mass-diffusion, electric-magnetic, and wave-constant energy over motive time.

$$E(d) = \frac{m_x D}{t_t} + \frac{p_z q_y}{t_t} + \frac{h_p w}{t_t}$$

Since the dynamic energy relationship is comprehensive, some of the terms may approach zero in selective or specialty analysis. If one of the phenomena goes to zero, then the equation is dramatically simplified. For example, many mechanical problems can be solved with only the second manifestation term while the electromagnetic and wave manifestations go to near zero.

Light _____

The dynamic conversion combines all the forms of energy. Dynamic energy becomes the forcing function for the natural, physical system.

Real physical systems adhere to the law of Conservation of Energy. *There is nothing new under the sun.* The more traditional form states *the sum of the energy in a closed system is zero.*

$$\Sigma E = 0$$

Energy is neither created nor destroyed, it may only change form.

If there is no external influence, then the sum of the energy equation is zero. Without external influence, the system relationship is called a characteristic equation. At the boundary, the energy is light.

The diffusion of energy 'D' can act on a wave 'w', on electromagnetic phenomenon 'p q', and on a corpuscular mass 'm', depending on the perspective. The relationship is very dependent on frequency and time.

Characteristic equation of dynamic energy

$$0 = \frac{m_x D}{t_t} + \frac{p_z q_y}{t_t} + \frac{h_p w}{t_t}$$

$$-\frac{D}{t_t} = \frac{p_z q_y}{m_x t_t} + \frac{h_p w}{m_x t_t}$$

The rate of diffusion of energy and light is the interaction of mass, charge, magnetism, and waves.

The negative sign indicates direction. By adjusting the order of the vectors underlying the regents, the sign change is resolved. The dynamic relationship is for an independent source. The negative sign indicates the terms are derived rather than applied.

A closed system is one without exterior influence. The system will decay to a stable condition. This control response is from a negative feedback. Positive feedback causes growth, but this comes only from open systems with outside impetus.

Restructure_____

From the three manifestations in the dynamic energy equation, each value can be related to time. This creates a rate parameter.

$$\dot{m} = {}^m\!/t_t \qquad\qquad \text{mass flow}$$

$$\dot{D} = {}^D\!/t_t \qquad\qquad \text{diffusion rate}$$

$$v = {}^p\!/t_t \qquad\qquad \text{voltage}$$

$$i = {}^q\!/t_t \qquad\qquad \text{current}$$

$$f = {}^w\!/t_t \qquad\qquad \text{frequency}$$

$$E/w = {}^{h_P}\!/t_t \qquad\qquad \text{energy per cycle}$$

The system quantities are the basis for a generalized definition of regent energy.

Definition of Regent Energy
Regent energy is the product of a complementary regent 'n_c' on a basis regent or value over time.

$$E = \frac{n_c n}{t_t}$$

Motive-force is associated with regent 'n'. The equation of motive force energy is proposed.

Definition of motive force energy

Energy is the motive-force operating on the basis regent. Motive force is the complementary regent over time.

$$E = mf * n$$

Where,

$$mf = \frac{n_c}{t_t}$$

When the basis is poles, the complementary charge makes magneto-mf in amperes. When the basis is charge, the complementary pole makes electro-mf in volts. Similarly, when mass is the basis, the complementary diffusion makes gravity-mf in 'm^2/s^2'.

Static_____

Each of the regents operates along an internal axis. They can be represented with a subscript relating to the axis. They appear mutually orthogonal.

The static mass is compared to the origin along the 'x' axis. The static charge operates along the 'y' axis. The magnetic pole is along the 'z'. To maintain neutrality, each of these will form a closed path along a surface vector.

Regent	Internal	Surface
m	x	t
q	y	t
p	z	s

The relative orientation of the charge and magnetism has been confirmed. Both 'x' and 'y' are reference axes, since they intersect with the tangential latitude 't'. The reference reactions will be similar except delayed in time. Further research needs to be conducted to ascertain the definitive axis for the mass and waves.

Fields _____

Fields and fluids are the same phenomenon. Matter consists of three regents - mass 'm', charge 'q', and magnetism 'p'. In the fundamental perception, each of these occurs at a point. In fields, their effects are distributed over a curved space volume. The magnetism is then referred to as magnetic flux 'Φ'.

Field analysis simply implies having dynamic field volume in the numerator and the space reference vector in the denominator.

A recurrent concept is the dynamic volume gradient 'dvg'. The dynamic volume gradient is inherent in diffusion and is linked to mass.

$$dvg = \frac{b_{rs} \times d_t \cdot s_r}{s_r}$$

Field energy becomes the point or node energy multiplied by the distribution ratio, which keeps the units intact. The ratio is dynamic field distances necessary to create volume in the numerator over the corresponding system space volume. This is linked to electric-magnetic and wave energy.

$$dr = \frac{b_{rs}\, d_t\, s_r}{s_s s_t s_r} = \frac{b_{rs}\, d_t\, s_r}{V_r}$$

The same space reference vector 's_r' is intentionally in both the numerator and the denominator. In some issues, the gradient factor appears to cancel, which would dimensionally yield an area rather than volume.

For mass-diffusion manifestation, the distribution ratio is only the space reference vector. Diffusion already includes the other two surface volume terms.

The dynamic energy for field conditions incorporates the dynamic volume in the numerator and the space reference in the denominator in each manifestation. The diffusion is expanded to yield its area vectors, so the complete space-time relationship is apparent

Equation of dynamic energy, field model

$$E(d) = \frac{m_x}{t_t}\frac{b_{rs}d_t s_r}{t_r s_r} + \frac{p_z q_y}{t_t}\frac{b_{rs}d_t s_r}{s_s s_t s_r} + \frac{h_p w}{t_t}\frac{b_{rs}d_t s_r}{s_s s_t s_r}$$

The field operates on the surface of the volume. The axis of action is the space reference vector in the denominator.

The space reference vector replaces the conventional defining partial derivative that yields field definition, direction, and results.

Mass Diffusion Field _____

To illustrate the field model, a few fundamental concepts are presented. Although some of the concepts in space and time are known, they are necessarily added for completeness.

Mass: The first manifestation of the energy model describes the mass-diffusion field equation.

Law of mass-diffusion energy, field model
Mass-diffusion energy is the product of the mass and dynamic volume over divergent times and space reference vector.

$$E(d) = \frac{m_x}{t_t}\frac{b_{rs}d_t s_r}{t_r s_r}$$

Einstein's relationship as well as Newton's Force Law and all its elegance may be extracted from this one term. From the field model, pressure and flow disturbance as well as stress and strain can also be explained. One illustration, Newton's relationship, is shown.

Energy is force acting through a lever arm distance 'b_{rs}'. For point or node, lumped conditions, the mass is not distributed over a volume. So, the space reference vector cancels.

$$E = b_{rs} * F_r$$

Where,

$$F_r = \frac{m_x d_t}{t_r t_t}$$

For a closed system, each time causes a 90 degree shift. The first time shift is from the relative to the tangential axis. The second time shift is from the tangential back to the reference axis on the negative side, but moving in the positive direction. Then the acceleration is back on the reference axis.

$$a = \frac{d_t}{t_r t_t}\bigg|_{m=k}$$

So,

$$F_r = m_x a_x$$

As another strategic example, the energy per volume is mass over divergent times and the space reference vector. This is also the dynamic viscosity rate.

$$\frac{E}{V} = \frac{m_x}{t_r t_t s_r}$$

Electric Magnetic Field _____

E-M: The second manifestation of the dynamic energy equation provides the electric-magnetic field equation.

Law of electric-magnetic energy, field model
Electric-magnetic energy is the product of magnetism, charge, and dynamic volume over time and space volume.

$$E(d) = \frac{p_z q_y}{t_t} \frac{b_{rs} d_t s_r}{s_s s_t s_r}$$

For point or node analysis, the distance vectors are aligned and cancel effects. Then the manifestation provides the circuit laws for a closed system. These include Faraday's, Lenz', and Kirchhoff's.

When expanded in terms of distance, the dynamic field equations are described. From this one relationship, Poynting's vector and Maxwell's suite of equations can be extracted as shown in the next chapter.

Energy is force acting through a lever arm distance 'b_{rs}'. For conditions where the reference space vectors cancel, the force is dependent on the wavelength and the orthogonal area, which is the surface of a spheroid.

$$E = b_{rs} * F_r$$

So,

$$F_r = \frac{p_z q_y}{t_t} \frac{d_t s_r}{V_r} = \frac{p_z q_y}{t_t} \frac{d_t}{A_r}$$

In a uniform circular space, the area and volume would have the following average values.

$$A_r = s_s \times s_t$$

$$A_r = 4\pi s_r^2$$

Where,

$$V_r = s_s \times s_t \cdot s_r$$

$$V_r = 4\pi s_r^2 \cdot s_r$$

$$V_r = \frac{4}{3}\pi s_r^3$$

Wave Constant Field_____

Wave: From the third dynamic energy manifestation, wave theory is characterized.

Law of wave energy, field model
Wave energy is the product of Planck's universal constant,
number of waves (cycles), and dynamic volume over time and
space volume.

$$E(d) = \frac{h_p w}{t_t} \frac{b_{rs} d_t s_r}{s_s s_t s_r}$$

The reciprocal of time relationship is often referred to as frequency. A more precise definition incorporates the number of spins around the path.

$$f = \frac{w}{t_t}$$

By using discrete components, Planck's theory can be extracted. Since the universal constant 'h$_P$' is such a small number, the wave energy approaches zero for low frequencies, such as Newtonian analysis.

Waves are spherical and emanate in all directions from the source. The distance around the path is the wavelength 'd$_t$'. Motion travels along the tangential.

Moreover, the entire bundle moves through 'space' as a mass-diffusion rate and electric-magnetic group. The term that dominates determines which reference axis or ray 'r' that the action is on.

Every physical system and matter has energy. Therefore, there is an associated natural frequency.

Equation of natural frequency

$$\frac{w}{t_t} = \frac{m_x D}{h_p t_t} + \frac{p_z q_y}{h_p t_t}$$

For living organisms, this is most noted in the region of 1000 Hz.

The harmonic wave value 'w' can be ascertained independent of time.

Equation of wave diffusion

$$w = \frac{m_x D}{h_p} + \frac{p_z q_y}{h_p}$$

Although this relationship is in its elemental form, expand diffusion to demonstrate the wave regent relationship's dependence on lever ray 'b_{rs}', wavelength 'd_t', and reference seasonal time.

Equation of wave fundamentals

$$w = \frac{m_x b_{rs} d_t s_r}{h_p s_r} + \frac{p_z q_y}{h_p}$$

The relationship illustrates circular response 't' to an object or impulse dropped in a fluid of mass 'm'. It also shows the orthogonal ray 'r' of an electric-magnetic transmission. The ray angle will deflect at each transition interface.

Velocity of propagation depends on the lever ray distance and material properties. In circular motion, the wavelength obviously is related to the radius or force ray length.

$$u_{rt} = \frac{b_{rs}}{t_r}$$

This appears in contrast to electromagnetic velocity 'u_t', but it corresponds to mass velocity. Therefore, wave and mass velocity depends on the diffusion. Electromagnetic properties depend on the relative permeability and permittivity of the medium, which are determined by the diffusion rate.

Energy is force acting through a lever arm distance 'b_{rs}'. For conditions where the reference space vectors cancel, the force is dependent on the wavelength and the orthogonal area, which is the surface of a spheroid.

$$E = b_{rs} * F_r$$

So,

$$F_r = \frac{h_p w_r}{t_t} \frac{d_t s_r}{V_r} = \frac{h_p w_r}{t_t} \frac{d_t}{A_r}$$

Diffusion _____

Diffusion is the rate of area change in dynamic space. Diffusion rate is a velocity product of divergent time.

$$\frac{D_r}{t_t} = \frac{b_{rs} d_t \, s_r}{t_r t_t s_r}$$

$$= u_{rt} u_t$$

When operating toward the interface, the velocities move to parallel and approach the speed of light in a vacuum.

$$\frac{D_r}{t_t} = c^2 |_{interface}$$

Diffusion rate was previously identified as regent dependent.

$$-\frac{D}{t_t} = \frac{p_z q_y}{mt_t} + \frac{h_p w}{mt_t}$$

Applying the velocities to the diffusion, gives a relationship for determining how close motion will be to the speed of light.

$$c = \sqrt{\frac{p_z q_y}{mt_t} + \frac{h_p w}{mt_t}}$$

Obviously, the relationship can fit the sum of the squares form of a right triangle. That relationship is fundamental to relativistic physics, as shown below.

The mass-diffusion definition encompasses the familiar mass to energy conversion concept developed by Einstein. Moreover, the definition contains Newton's equations. It addition, it provides one of the components for the harmonic motion of string theory.

Relative Motion _____

Relativity is the process of comparing the motion between two items.

Imagine observation of motion from the surface back toward the origin. The motion appears as a line perpendicular to the reference line. Then, the perception of velocity is a square root of the difference of the squares.

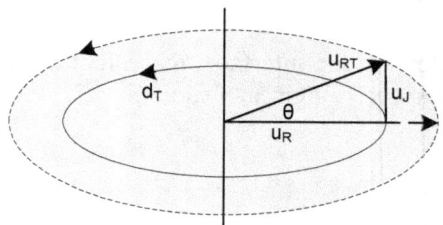

Figure 2 – Velocity Vectors

The interface velocity 'u_{rt}' is the arc motion at the tip of the ray. The forward velocity component 'u_j' is along the imaginary and is a projection of 'u_{rt}'. When comparing arc motion to a point at the origin, the relative or line-of-sight velocity component 'u_r' along the reference axis can be calculated.

$$\frac{u_r}{u_{rt}} = \frac{\sqrt{u_{rt}^2 - u_j^2}}{u_{rt}}$$

Or,

$$\cos \theta_{rt} = \frac{\sqrt{u_{rt}^2 - u_j^2}}{u_{rt}}$$

The cosine term becomes the Lorentz factor when the radius velocity 'u_{rt}' at the interface is the speed of light. Use the known rest mass 'm_0' along the reference axis. With this definition, the motive mass 'm_{rs}', as perceived moving along the arc at the ray, can be determined.

$$\frac{m_0}{m_{rs}} = \frac{u_r}{u_{rt}}$$

Then,

$$m_{rs} = \frac{m_0}{\cos \theta_{rt}}$$

$$= \frac{m_0}{\sqrt{c^2 - u_j^2/c}}$$

The key component to the discussion is the elegance of relativity as well as the simplicity of Newtonian motion are both incorporated in the curve space and divergent time relationships of the dynamic spheroidal coordinate system. The relativity concepts are as fundamental as the lever ray vector 'b_{rs}' relationship to the reference axis 'b_r'.

Review_____

The equation of dynamic energy, node model is the fundamental unified triad energy relationship. Dynamic energy is the sum of mass-diffusion, electric-magnetic, and wave-constant energy over time.

Equation of dynamic energy, node model

$$E(d) = \frac{m_x D}{t_t} + \frac{p_z q_y}{t_t} + \frac{h_p w}{t_t}$$

Field analysis simply implies having dynamic volume in the numerator and the space reference vector in the denominator.

Field energy becomes the point or node energy multiplied by the distribution ratio, which keeps the units intact. The ratio is dynamic field distances necessary to create volume in the numerator over the corresponding system space vectors.

Equation of dynamic energy, field model

$$E(d) = \frac{m_x}{t_t} \frac{b_{rs} d_t s_r}{t_r s_r} + \frac{p_z q_y}{t_t} \frac{b_{rs} d_t s_r}{s_s s_t s_r} + \frac{h_p w}{t_t} \frac{b_{rs} d_t s_r}{s_s s_t s_r}$$

The field operates on the surface of the volume. The axis of action is the space reference vector in the denominator. The space reference vector 's$_r$' replaces the conventional defining partial derivative that yields field definition, direction, and results.

$$\Leftarrow \Uparrow \Rightarrow$$

4

ELECTROMAGNETIC FIELDS IN ONE CURVED SPACE EQUATION

Thought
Electric and magnetic are two-thirds of matter,
but they cannot be seen, heard, smelled, or tasted.

Abstract _____

Electromagnetic interaction is illustrated with one equation. Curved space and divergence of time provide a model that covers circuits as well as fields. Definitions of measured parameters as well as Kirchhoff's Laws and Maxwell's suite are included in the relationship. By judicious application of these principles, the driving energy for every electrical, magnetic and electromagnetic field problem in high energy physics and engineering can be resolved.

The previous chapter correlates the dynamic mass, waves, and light. This chapter uses the unique vector characteristic to investigate electric-magnetic interactions in one equation. The next chapter in the trilogy addresses the spin on curved space and divergence of time.

Introduction _____

Energy is the concept for conversion between physical systems. Energy or light consists of three components - space, time, and matter. Space has three measures - motion, ray, and volume. Volume has three dimensions – tangential, standing, and reference.

Time has three manifestations - constant called aeon, cyclic motion called chronos, and seasonal reference called kairos. Matter is three regents - mass, charge, and magnetism. This perception of natural physical systems provides a principle for research models.

What is the matter?   _____

Matter consists of three regents - mass 'm', charge 'q', and magnetism 'p'. Each regent has its own energy domain. In the fundamental perception, each of these occurs at a point or node. In fields, their effects are distributed over a volume. The magnetism is then referred to as magnetic flux 'Φ'.

Mass yields mechanical or gravitational energy, charge yields electric energy, and magnetic poles yield magnetic energy.

Electromagnetic influence is two-thirds of matter, but it cannot be seen, heard, smelled, or tasted. Nevertheless, its energy is very real and the node form can be completely expressed in one simple law of three variables.

Law of Electric-Magnetic Energy, node form
Electromagnetic energy is the product of magnetism, p, and charge, q, over motive time.

$$E = \frac{p_z q_y}{t_t}$$

Energy relationships are changing values related to a reference. Time is a denominator function creating a rate. The resulting energy is an instantaneous peak. Therefore, calculus representation is superfluous. However, derived concepts and average values may depend on the calculus.

The subscript shows the relative axis on which the activity occurs.

Curved Space _____

Curved space analysis using a spheroidal coordinate system was introduced in earlier chapters and will be expanded in the next. The figure is included here for reference to the subscripts.

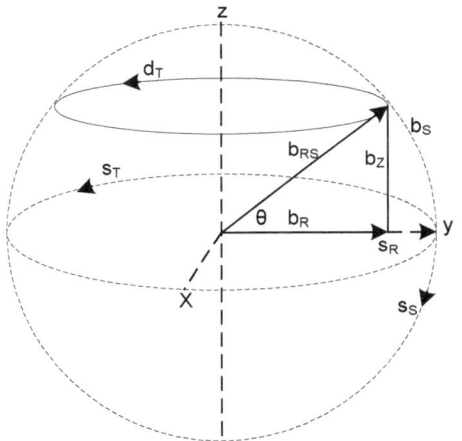

Figure 1 – Curved Space Vectors

Notice that the internal axes are a projection of the surface axes. For example, 's_y' is a projection of 'd_t', and 's_z' is a projection of 's_s'. The general reference subscript 'r' represents the 'x' or 'y' internal axis, depending on the system.

Node _____

From the one electromagnetic energy relationship, all electrical and magnetic circuit concepts at a point or node can be directly derived. The three definitions of electrical measurements are embedded in the energy equation.

The definition of voltage is Faraday's Law. Voltage is the change in magnetic flux with time, when charge is constant or not changing. When it is a driving source, voltage is called electromotive force, *emf* or *e*.

$$v = \frac{p_z}{t_t}\bigg|_{q=k}$$

Lenz's law, which gives the direction of induced voltage, is included in the subscript direction.

The definition of current is the change in charge with time, when magnetism is constant or not changing. When it is a driving source current is called magnetomotive force, *mmf*.

$$i = \frac{q_y}{t_t}\bigg|_{p=k}$$

The definition of frequency is the reciprocal of time.

$$f = \frac{1}{t_t}$$

Using the definitions, electromagnetic energy can be segregated into other forms. Electrical energy is the product of voltage and charge.

$$E_{electric} = vq$$

Magnetic energy is the product of current and magnetic flux.

$$E_{magnetic} = ip$$

Obviously, there is a very close relationship between charge and magnetism. One does not depend on the other. They can exist independently as seen in the node form. However, when there is motion represented by the time, then one influences the other. The motion of charge and magnetism is the basis of electric machines and electromagnetic fields.

Conservation _____

Real physical systems adhere to the law of Conservation of Energy. *There is nothing new under the sun.* The more traditional form states *the sum of the energy in a closed system is zero.*

$$\Sigma E = 0$$

Energy is neither created nor destroyed, it may only change form. When applying the conservation of energy relationship to the node form of the electromagnetic energy equation, an entire paradigm is developed and other conservation laws necessarily exist. Since charge, magnetism, and time cannot convert to the other, conservation applies to each item individually.

Conservation of charge states the sum of the charge is zero. Charge is discrete and is an integral multiple of the charge on an electron or proton.

Conservation of magnetism states the sum of the magnetic flux is zero. Magnetic poles always exist in a balanced pair or dipole with a north and south pole, since magnetic lines must make a closed loop. The poles are where magnetism leaves the magnetic to form a field.

Conservation of time and frequency implies the sum of frequency is zero. Alternatively, time is conserved. Conservation of time and frequency is directly related to Planck.

$$E = hf$$

These conservation relationships are implicit in the electromagnetic energy expression when the constraint of conservation of energy is imposed.

Circuit Analysis and Power _____

Circuit analysis is often described using two laws developed by Kirchhoff. Both of these laws are imbedded in the electric-magnetic energy law. First, apply conservation to the relationship. This sets the sum of the energy equal to zero. Next, hold one term constant, then the sum of the changing term is zero.

Kirchhoff's current law (KCL) can be stated: *When the magnetic flux is constant, the sum of the current at a node is zero.*

$$\Sigma E = 0$$

$$\Sigma \frac{p_z q_y}{t_t} = 0$$

Then,

$$\Sigma i = 0|_{p=k}$$

Similarly, Kirchhoff's voltage law (KVL) can be stated: *When the charge is constant, the sum of the voltage around a path is zero.*

$$\Sigma \frac{p_z q_y}{t_t} = 0$$

Then,

$$\Sigma v = 0|_{q=k}$$

Energy can have either voltage or current, but not both. Calculations that have both will involve power. Power is the energy over time. Power '$\mathcal{P}$' imposes another time divergence on the electromagnetic energy.

$$\mathcal{P} = \frac{E}{t_r}$$

$$\mathcal{P} = \frac{p_z q_y}{t_r t_t}$$

$$\mathcal{P} = vi$$

Fields _____

Fields and fluids are the same phenomenon. Matter consists of three regents. In the fundamental perception, each of these occurs at a point. In fields, their effects are distributed over a curved space volume.

Field analysis simply implies having dynamic volume in the numerator and the space reference vector in the denominator.

A recurrent concept is the dynamic volume gradient 'dvg'. The dynamic volume gradient is inherent in diffusion.

$$dvg = \frac{b_{rs} \times d_t \cdot s_r}{s_r}$$

Field energy becomes the point or node energy multiplied by the distribution ratio, which keeps the units intact. The ratio is dynamic distances necessary to create volume in the numerator over the corresponding system space distances.

$$dr = \frac{b_{rs}\, d_t\, s_r}{s_s s_t s_r} = \frac{b_{rs}\, d_t\, s_r}{V_r}$$

Note the same space reference vector 's_r', is in both the numerator and the denominator. As a result, the vectors can cancel and many physical perceptions can be described in terms of area.

> *Law of Electric-Magnetic Energy, field form*
> *Electric-magnetic energy is the product of magnetism, p, and charge, q, over time, multiplied by the lever ray, motion, and space reference vectors over the system space volume.*

$$E = \frac{p_z q_y}{t_t} \frac{b_{ys} d_t s_y}{s_s s_t s_y}$$

The single, unified electromagnetic field relationship includes the product of electric intensity and density, the product of magnetic intensity and density, the Poynting vector, and all four of Maxwell's equations. These individual concepts will be developed below.

Force _____

Energy is force through a distance. The circular or star operator provides both linear and rotational energy when using the ray 'b_{ys}'.

$$E = b_{ys} * F_y$$

Comparing this relationship with the field electromagnetic energy, then force includes all the terms except the ray 'b_{rs}'.

Because of the vector product associated with 'b_{ys}', the energy relationship includes linear or dot-product energy along the reference axis 'y', and rotational or cross-product energy from torque 'T' around a curved, closed, tangential path 't'.

$$E = b_{rs} * F_r$$
$$= b_r \cdot F_r + (T_t \theta_{rs})_t$$

Field Intensity and Density _____

Field properties are measured relative to a dimension, either distance or area.
- *Intensity* is a measured parameter - voltage or current - over a distance.
- *Field intensity* is a dynamic function since it depends on time.
- *Density* is matter over area.
- *Energy* is the product of intensity, density, and volume.

These electrical and magnetic field equations can be extracted from the energy field relationship above.

The electric intensity '$\mathcal{E}$' is voltage over the measurement path, 's_t'.

$$\mathcal{E}_t = \frac{p_z}{t_t s_t}$$

The magnetic intensity '$\mathcal{H}$' is current within a closed path 's_s' as seen in Figure 2.

$$\mathcal{H}_s = \frac{q_y}{t_t s_s}$$

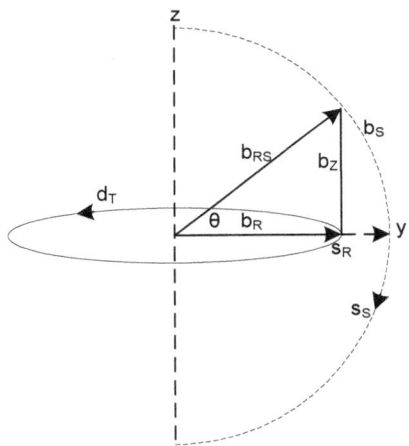

Figure 2 – Closed Path

Field density is a static function. Since density is independent of time, and time is a surface function, then density must be spread over an internal area. The electric density '$\mathcal{D}$' is charge based.

$$\mathcal{D}_y = \frac{q_y}{A_y}$$

The magnetic density '$\mathcal{B}$' is magnetic strength spread over the area the lines penetrate.

$$\mathcal{B}_z = \frac{p_z}{A_z}$$

These four relationships are the foundation of electro-magnetic field analyses. They are directly related to Maxwell's suite, but are contained within and extracted from the single, unified, electromagnetic field relationship.

In addition, consider circuit analysis. Rather than being a point used in most circuit analysis, current 'i_t' is actually spread around the perimeter of the conductor as a skin effect. The perimeter is determined by the cross-sectional area 'A_t' as seen in Figure 3. The current density '$\mathcal{J}_t$' is the current over the area. The semicircle of Figure 2 is collapsed into a tube.

$$\mathcal{J}_t = \frac{q_y}{t_t A_y}$$

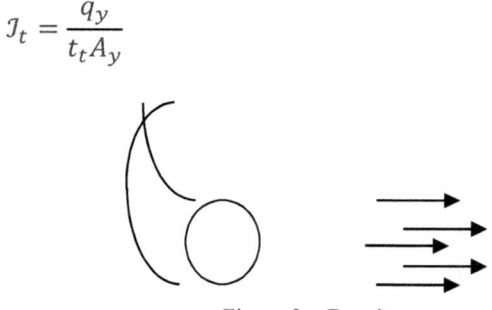

Figure 3 – Density

If current is dispersed, it is logical to expect the underlying charge 'q_y' to occupy more than a point. Charge spread over the volume is called charge density 'ρ'.

$$\rho_y = \frac{q_y}{V_y}$$

The very basic, simple, unified electric-magnetic energy relationship provides any circuit concept. A related equation with volume permits the solution of all the rest of the problems that involve distributed energy or fields.

Diffusion

Using dynamic space and divergent time, diffusion was defined in a previous chapter. The rate of diffusion is a velocity product.

$$\frac{D_r}{t_t} = \frac{b_{rs} d_t}{t_r t_t} \frac{s_r}{s_r}$$

$$\frac{D_r}{t_t} = u_{rt}u_t$$

The vectors are seen in the diagram.

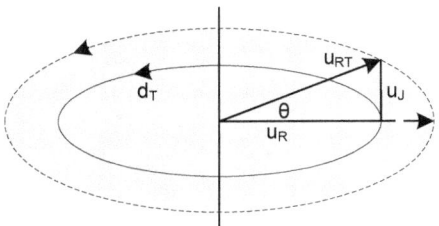

Figure 4 – Velocity Vectors

The boundary of the velocity product depends on the electromagnetic property of the material. The electric or charge parameter is permittivity 'ε'. The magnetic or pole parameter is permeability 'μ'.

$$u_{rt}u_t = \frac{1}{\mu\varepsilon}$$

Both parameters have a constant value in a vacuum '0' and a relative reference value 'r' for each material environment.

$$\mu = \mu_r\mu_0$$
$$\varepsilon = \varepsilon_r\varepsilon_0$$

Then the velocity depends on the reference values and vacuum values.

$$u_{rt}u_t = \frac{1}{(\mu_r\varepsilon_r)(\mu_0\varepsilon_0)}$$

In a simple vacuum, without electromagnetic influence, the reference parameters equal 1. When the rate is operating toward the interface, the velocities are parallel. Then, the velocity observed from near the interface approaches the speed of light. At that interval, the rate of diffusion is the speed squared.

$$u_{rt}u_t = c^2|_{interface}$$

The speed of light is then dependent on the diffusion velocities in a vacuum.

$$c = c_r c_0$$

Where,

$$c_0 = \frac{1}{\sqrt{\mu_0 \varepsilon_0}}\Bigg|_{vacuum}$$

The velocity product gives a relationship for determining how close motion will be to the speed of light.

The limit on speed is the relative permeability and permittivity of the medium.

Summary _____

Electric-magnetic energy is made up of electrical charges and magnetic poles moving in cyclic motive time frame.

$$E = \frac{p_z q_y}{t_t}$$

The circuit or rotational motion operates on a spheroid. Simply by maintaining the directional orientation, it is possible to identify all field problems from one very simple vector relationship.

$$E = \frac{p_z q_y}{t_t} \frac{b_{ys} d_t s_y}{s_s s_t s_y}$$

One equation can describe all the electromagnetic analysis. The complete model includes fields and diffusion in curved space. When the distances are resolved, the relationship simplifies to a circuit problem as shown in the node equation.

By judicious definition of the star operator, a simple field energy equation incorporates both lineal and rotational motion. Relative as well as Newtonian motion is contained in the relationships.

Application of the reference space vector 's$_y$' maintains orientation, describes Maxwell's suite, and makes the del operator unnecessary.

Annex _____

The suite of equations developed by Maxwell contains four relationships.

$$\nabla \times \mathcal{E} = -\frac{d\mathcal{B}}{dt} \qquad\qquad Volt/m^2$$

$$\nabla \times \mathcal{H} = \mathcal{J} + \frac{d\mathcal{D}}{dt} \qquad\qquad Amp/m^2$$

$$\nabla \cdot \mathcal{D} = \rho \qquad\qquad Coulomb/m^3$$

$$\nabla \cdot \mathcal{B} = 0$$

Using the common internal, space reference vector '1/s$_y$', rather than the del, the suite of four equations can be calculated from the single, unified, electric-magnetic energy field relationship.

$$E = \frac{p_z q_y}{t_t} \frac{b_{ys} d_t s_y}{s_s s_t s_y}$$

First, the intensity or density relationship will be shown as defined above. Next, to obtain volumetric space terms, both sides of the equation will be multiplied by the inverse of the reference vector along the y-axis, '1/s$_y$'. Then the subsequent equations manipulate the vector algebra. The result is a relationship that is equivalent to one of the del equations of Maxwell.

This simple process uses a unified, electromagnetic equation with a vector along an axis. This eliminates the complex calculus of Maxwell in exchange for a simple algebra operation.

Intensity: The distances used in the dynamic or intensity relationships are on to the external, curved space axes 's$_s$, s$_t$, s$_y$'. These surface, orthogonal vectors inherently contain the cross product of the del '∇'.

Multiplying the intensity by the *1/ s$_y$* vector provides a relationship equivalent to a Maxwell term. The vector in the reference direction 's$_y$' multiplied by a vector on the surface yields an area in the other surface direction.

1.   Equation of electric intensity

$$\mathcal{E}_t = \frac{p_z}{t_r s_t} \qquad \text{Volt/m}$$

Apply the reference vector *1/ s$_y$*.

$$\left(\frac{1}{s_y}\right)\mathcal{E}_t = \frac{p_z}{t_r s_y s_t} \qquad \text{Volt/m}^2$$

$$= \frac{p_z}{t_r A_{-s}}$$

$$= \left[\frac{\mathcal{B}_z}{t_r}\right]_{-s} = -\left[\frac{\mathcal{B}_z}{t_r}\right]_s$$

$$= \nabla \times \mathcal{E}$$

2a. Equation of magnetic intensity

$$\mathcal{H}_s = \frac{q_y}{t_t s_s} \qquad \text{Amp/m}$$

Apply the reference vector *1/ s$_y$*.

$$\left(\frac{1}{s_y}\right)\mathcal{H}_s = \frac{q_y}{t_t s_y s_s} \qquad Amp/m^2$$

$$= \frac{i}{A_t}$$

$$= \mathcal{J}$$

$$= \nabla \times \mathcal{H}$$

2b. Equation of charge density

$$\mathcal{D}_y = \frac{q_y}{A_y} \qquad Cb/m^2$$

Apply the frequency $1/t_t$.

$$\frac{\mathcal{D}_y}{t_t} = \frac{q_y}{t_t A_y} \qquad Amp/m^2$$

$$= \frac{i}{A_y}$$

$$= \mathcal{J}$$

$$= \nabla \times \mathcal{H}$$

Density: The distances in the static or density relationships are relative to the internal, reference axes 's_x, s_y, s_z'. The internal axes inherently contain the dot product of the del '∇'.

Multiplying the density by the $1/s_y$ vector provides a relationship equivalent to a Maxwell term.

3. Equation of electric density

$$\mathcal{D}_y = \frac{q_y}{A_y} \qquad Cb/m^2$$

Multiply by the $1/s_y$ vector.

$$\left(\frac{1}{s_y}\right)\boldsymbol{\mathcal{D}}_y = \frac{q_y}{s_y A_y}$$

$$\qquad Cb/m^3$$

$$= \frac{q_y}{V_y}$$

$$= \rho_y$$

$$= \nabla \bullet \boldsymbol{\mathcal{D}}$$

4. Equation of magnetic density

$$\boldsymbol{\mathcal{B}}_z = \frac{p_z}{A_z}$$

$$\qquad Wb/m^2$$

Multiply by the $1/s_y$.

$$\left(\frac{1}{s_y}\right)\boldsymbol{\mathcal{B}}_z = \frac{p_z}{s_y A_z}$$

$$\qquad Wb/m^3$$

$$= 0$$

$$= \nabla \bullet \boldsymbol{\mathcal{B}}$$

The vector in the reference direction 's_y' multiplied by the plane area in the direction of the regent yields a volume. In the magnetic equation, the radial and the plane area are in different directions. Hence, the result of a dot product in two different directions does not exist.

It is fascinating that all the action is on the reference axis 's_y'. The reference vector replaces the complexity of the partial derivatives of the del. However, it is the understanding of physical relationships that make the unified, electric-magnetic equations possible.

$$\Leftarrow \Uparrow \Rightarrow$$

5

SPIN ON CURVED SPACE COORDINATES AND DIVERGENCE OF TIME

Thought
*Time analysis varies from independence
through time to time divergence.*
MOD

Abstract _____

Physical systems operate in a curved space. An alternative curved coordinate system resolves both gravitational and electromagnetic problems. The effects of time on curved space create a divergence that redefines the position regent. By judicious application of these principles, the driving energy for natural, physical systems can be resolved. A method of mathematics is defined and structured.

The principles are presented in this chapter. The previous chapter used these unique vector characteristics to investigate electric-magnetic interactions in one equation. The first of the trilogy correlated the dynamic mass, waves, and light.

The curved space and divergent time has been extended to other laws such as static energy and thermal energy.

Introduction _____

Physical systems operate in curved space. This has traditionally created difficulty because most analysis uses a rectangular Cartesian coordinate system. The complexity of Maxwell's suite arises from the calculus on a rectangular system.

Similarly, Einstein's relativity has conceptual challenges because of the coordinate system and the number of terms necessary to describe an adjacent, relative position using tensors.

In some systems, improvement is obtained using polar notation, particularly for electrical circuits and rotational dynamics.

An alternative coordinate system greatly improves conceptual illustration of these physical networks. The development of any analysis is highly dependent on the transportation system used as a reference. Galileo did his work based on the relatively slow movement of sailing ships. Einstein related his elegant concepts to the faster train.

Imagine the movement of aircraft and near space vehicles. The navigation is relative to latitude, longitude, and altitude. A modification to traditional navigation provides an elegant, general tool for all motion whether micro or macro and whether gravitational or electromagnetic. With this curve space reference as the basis of a spheroidal coordinate system, it is much easier to define electromagnetic and relative motion.

Curved Space_____

Space and motion in our perspective is a three-dimensional phenomenon requiring three volume coordinates. In addition, three different types of space measurements are required. This perception of natural physical systems provides a principle for research models.

All calculations are changes in values of variables from a reference. The relations represent the instantaneous and peak values. Therefore, calculus notation is superfluous. However, derived concepts and average values may depend on the calculus.

Space is a three-dimensional spheroid with measurements on the surface. The three types of curved space distance or dimensions are motion, torque lever ray or rotational, and space reference.

Motion distance or wavelength 'd_t' is measured as a latitude line tangential around a sphere. The lever arm is a ray 'b_{rs}' that is projected onto the reference or relative axis 'b_r' and the longitude 'b_s'. The motion distance passes through the end point of the ray. The ray is not along an axis and the length will vary with the contour.

Space distance 's' has three components that encompass a distance along a latitude or tangential 's_t', a distance up a longitude or standing coordinate 's_s', and a distance along a reference 's_r'. The reference may be oriented along an internal Cartesian coordinate 'x, y, z' to provide a transform into that reference system.

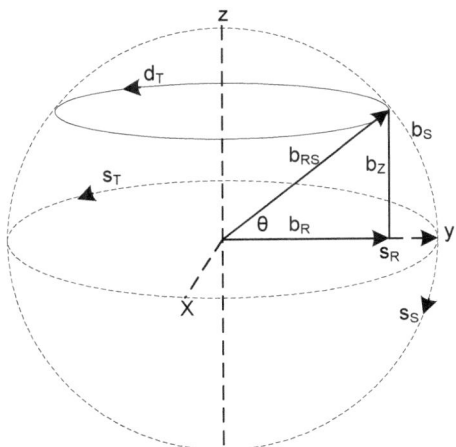

Figure 1 – Curved Space Vectors

In a limited region, a rectangular 'x, y, z' assumption is acceptable and is used for linear analysis. Vectors relative to the surface axes are projected onto the rectangular axes.

The reference axis 'b_r' appears to be a straight line in the close context of analysis. However, it is a curved tangential path similar to 'd_t' on a larger system analysis.

Imagine a satellite orbiting relative to a center nucleus or sun. Spheroid analysis is the action on the surface of the satellite. The reference axis and time, in a short term, appear straight. In a larger global view, they are a curve on the surface or orbit about the nucleus system. Relative motion on the satellite is contained in the subscripts.

The volume does not have to be uniform. The volume can be represented as the average area of the surface of a Riemann sphere and the dot product of the volume distance.

Manipulation _____

An expansion of the conventional cross and dot product is used for calculations. Components in the same direction have the same subscript, and their products are in the same direction. Components that are orthogonal have different subscripts and the product is in the third direction.

The conventional internal area and volume are noted in the Cartesian system. Generally, volume is assumed to not have direction, but the subscript will be maintained to indicate the reference axis of development.

$$A_y = s_z \times s_x$$
$$V_y = A_y \bullet s_y$$

Similarly, a curved space has a solid area and resulting spheroidal volume. These values are on the surface of the curved system.

$$A_y = b_s \times d_t$$
$$V_y = A_y \bullet s_y$$

Image a tennis ball that is flattened on a side. The point represented by the ray vector 'b_{rs}' changes, but the system surface area does not change. The spheroidal system represents any curved space and is not required to be uniform in shape and indeed, it will seldom be.

In a uniform circular space, the area and volume would have the following average values.

$$A_r = s_s \times s_t$$

$$A_r = 4\pi s_r^2$$

Where,

$$V_r = s_s \times s_t \cdot s_r$$

$$V_r = 4\pi s_r^2 \cdot s_r$$

$$V_r = \frac{4}{3}\pi s_r^3$$

Star Product _____

The ray 'b$_{rs}$' is projected onto the reference axis 'b$_r$' and the longitude or standing axis 'b$_s$'. Therefore, it has both an inline and an orthogonal arc component. To represent both actions, a new operation for a circuital or star product is necessary.

Circuital Law
The star product is the sum of the linear dot product and the curved cross product.

$$
\begin{aligned}
b_{rs} * e^{j\theta} &= b_{rs} * 1_r \\
&= \left(b_{rs} \bullet 1_r\right)_r + \left(\theta_{rs} b_{rs} \times 1_r\right)_t \\
&= \left(b_{rs} \bullet 1_r\right)_r + \left(b_s \times 1_r\right)_t \\
&= \left(b_{rs} 1_r \cos\theta_{rs}\right)_r + \left(b_s 1_r \sin\theta_{rs}\right)_t \\
&= \left(b_r 1_r\right)_r + \left(b_j 1_r \theta_{rs}\right)_t
\end{aligned}
$$

For simple rectangular systems, a special case of the circuital law is Euler's relationship, which describes Euclidean space. Euler tracks a point on a circle, but does not provide the distance about the circle.

The circuital operator uses a segment with an arc 'b_s'. In contrast, Euler uses a right triangle with an opposite side length 'b_j'. The opposite side only contains a projection of the arc, not the complete arc. Therefore, the corresponding Euler relationship is a subset of the star product.

$$b_{rj}e^{j\theta} = \left(b_{rs} \bullet 1_r\right)_r + \left(b_{rs} \times 1_r\right)_x$$
$$= \left(b_{rs}1_r \cos\theta_{rs}\right)_r + \left(b_{rs}1_r \sin\theta_{rs}\right)_x$$
$$= \left(b_r 1_r\right)_r + \left(b_j 1_r\right)_x$$

The length of the resultant radius 'b_{rs}' is equivalent to the hypotenuse 'b_{rj}'. Different subscripts are used to indicate different projections. The distance along the reference axis is 'b_r'; the arc length is 'b_s'; while the vertical projection is 'b_j' and the product direction 'x' is orthogonal to 'j' and 'r'.

Since much of the motion of dynamic energy is around a curved or closed path near the surface, the concept of a star product and field energy is natural.

Fields _____

Fields and fluids are the same phenomenon. Matter consists of three regents - mass 'm', charge 'q', and magnetism 'p'. Each regent has its own energy domain. In the fundamental perception, each of these occurs at a point. In fields, their effects are distributed over a curved space volume.

Field analysis simply implies having dynamic volume in the numerator and the space reference vector in the denominator. A recurrent concept is the dynamic volume gradient 'dvg'. The dynamic volume gradient is inherent in diffusion.

$$dvg = \frac{b_{rs} \times d_t \cdot s_r}{s_r}$$

The space reference is very intentionally maintained in both the numerator and denominator. It is necessary in order to have volume in the numerator and it is necessary in the denominator to identify fluids and fields problems. In some issues, the gradient factor appears to cancel, which would dimensionally yield an area rather than volume. As a result, many physical perceptions can be described in terms of area.

Field energy becomes the point or node energy multiplied by the distribution ratio, which keeps the units intact. The ratio is dynamic distances necessary to create volume in the numerator over the corresponding system space distances.

$$dr = \frac{b_{rs}\, d_t s_r}{s_s s_t s_r} = \frac{b_{rs}\, d_t s_r}{V_r}$$

Field energy encompasses both lineal and rotational energy and motion in one term, the circuital energy law.

Circuital Energy Law
When the star product is applied to the lever ray vector 'b_{rs}', and reference force 'F_r', the energy relationship includes linear or dot product energy along the reference axis and rotational energy from torque 'T' around a curved, closed, tangential path.

$$E = b_{rs} * F_r$$
$$= \left(b_{rs} \bullet F_r\right)_r + \left(\theta_{rs} b_{rs} \times F_r\right)_t$$
$$= \left(b_r F_r\right)_r + \left(T_t \theta_{rs}\right)_t$$

Time

Energy or light consists of three components – matter, space, and time. On a fundamental perception, time is reasoned by a person to be a sequential, linear event derived from when the person was born. In the paradox of time, Einstein illustrated that time is compressible.

A paradox implies there is an inadequately described perception which causes seeming contradictions. Some correlate time as a fourth dimension, but still with an inadequate description of performance.

Time is overlaid on the spheroidal coordinate system. Time associated with motion occurs along the surface tangential 't' axis. The second time is on a reference axis 'r'.

In a realized field, these can be orthogonal. As the times diverge, the projection of the other time appears more compressed. The third time predicted by the triad principle is fixed or unchanging.

For dynamic motion relationships, time is a denominator factor and forms a rate.

Delta conditions are without time change. Therefore, the time factor is '1'. Velocity conditions have a single time change '$1/t_t$'. Acceleration conditions have a second time change '$1/t_t t_r$'. The three times correspond to the second order definition of physical systems.

The time representation can be totally different realizations. *Aeon* is the concept for time placement with a value of one. It represents unchanging conditions for very long periods. *Chronos* has the subscript 't' for time associated with tangent or surface vectors. This concept is cyclic or repetitive as a clock. *Kairos* has the subscript 'r' for time associated with vectors along the reference. This representation appears linear in short intervals, but is actually seasonally cyclic.

In a curved system, time appears as a positional modifier, 't_t' on the path 'd_t' and as an expansion positional modifier, 't_r' on the lever arm 'b_{rs}'. Volumetric space 's' has unchanging time '$t=1$' associated with its values.

Previous chapters addressed the time effect on matter. The curved space time divergence is the topic of this treatise.

Velocity _____

The velocity is the rate of motion in a direction. Because of the two types of variable time, there are two types of velocity. The curved, closed space, or tangential velocity is the rate of change in the tangential distance.

$$u_t = \frac{d_t}{t_t}$$

The expansion or ray velocity is the rate of change in the lever over time.

$$u_{rt} = \frac{b_{rs}}{t_r}$$

Because of the positional performance of the lever, the velocity direction can vary from the reference axis 'v_r' to the tangential 'v_t'. The result can be from divergent to parallel velocities when related to the motion velocity 'u_t'.

Speed Limit_____

The speed of light binds time to motive distance.

$$d_t = ct_t$$

The time for one cycle or spin is proportional to the distance or length of the cycle 'd_t'. The speed of light is the proportionality constant.

The speed of light 'c' is a constant circle constraint that forms a spheroidal boundary interface on velocity. It is a constraint 'gravity' that holds the velocity limit to near the interface at the edge of the perceived three-dimensional realm.

Although cycles are generally considered as dimensionless, a discrete integer 'w' is used to count the number of waves, cycles, revolutions, spins, or vibrations.

$$\frac{d_t}{w} = c\frac{t_t}{w}$$

The quantity distance per cycle is wavelength 'λ' represented by 'd_t'. The boundary velocity is the speed of light.

$$\lim u_t = c$$

The reciprocal of time relationship is often referred to as frequency. A more precise definition incorporates the number of spins around the path.

$$f = \frac{w}{t_t}$$

Electromagnetics operate on the constant velocity circle at a wavelength distance. In Newtonian physics, mass operates in a small fraction of wavelength and time.

Imagine the surface of the earth. There is a band about the surface where life exists and people operate. Similarly, the constant circle is where the speed boundary exists, but there may be small variations from the nominal value.

The earth has a circumference of approximately 40×10^6 meters. Therefore, the time to traverse one wavelength is 0.133 seconds. So, the natural frequency of the earth is about 7.5 Hz.

$$t_t = \frac{d_t}{c} = \frac{40 \times 10^6}{3 \times 10^8} = 0.1333\,s$$

$$f = \frac{w}{t_t} = \frac{1}{0.133\,s} = 7.5\,Hz$$

Closed Path _____

Motion of a regent traverses a complete path. The regent begins in a static position along a reference axis 'r'. With time, the regent encounters velocity motion with a phase shift to the real, tangential 't' axis. The next time shift results in acceleration back to the reference, but coming from the negative side moving in the positive direction. Each time causes a phase shift of 90^0.

The length of the surface path is projected on the internal axis perpendicular to the starting reference axis.

$$d_t = 2\pi b_r$$

Angular frequency 'ω' arises in rotational or cyclic motion. The angle 'θ' traverses one complete cycle, circle, or revolution during the time sequence.

$$\omega = \theta \frac{w}{t_t} = 2\pi \frac{w}{t_t} = 2\pi f$$

Diffusion _____

Fields are distributed through curved space-time. Diffusion is the rate of dissipation to achieve a concentration. Diffusion is inherent to mass and may apply to any physical item whether electric-magnetic, waves, or light. Kinematic viscosity is related to diffusion.

Diffusion space is defined by the motion and ray vectors. The space reference vector 's_r', is eliminated in the numerator and denominator of mass diffusion field energy.

$$A_{-sr} = b_{rs} d_t$$

Dissipation is a velocity. The first application of time gives diffusion in the direction of the reference. Diffusion is the rate of area change.

Law of Diffusion
Diffusion is the product of lever ray, motion distance vector, and space reference vector over time and space reference vector.

$$D = \frac{b_{rs} \times d_t \cdot s_r}{t_r \, s_r}$$

The second, divergent time operating on the vectors yields rate of diffusion which is a velocity product.

$$\frac{D}{t_t} = \frac{b_{rs} \times d_t \cdot s_r}{t_r \, t_t \, s_r}$$

$$= \frac{b_{rs} \times d_t}{t_r \, t_t} \frac{s_r}{s_r} = u_{rt} u_t$$

When the rate is operating toward the interface, the velocities become parallel 'u_t'. Then, the velocity observed from near the interface approaches the speed of light. At that interval, the rate of diffusion is the speed squared.

$$\frac{D_r}{t_t} = c^2|_{interface}$$

The speed of light is then dependent on the diffusion velocities in a vacuum. Applying the parallel velocities to the calculation for diffusion, gives a relationship for determining how close motion will be to the speed of light.

Relative Motion

Relativity is the process of comparing the motion between two items. Consider the figures with an angle at the origin 'θ_{rs}', a ray 'b_{rs}', and an arc 'b_s'. A right triangle is created from the projection of the ray on the reference axis 'b_r' and the projection of the arc on an axis orthogonal to the reference 'b_j'.

Engineering calculations are generally performed at the origin. This is called the laboratory perspective. However, the actual motion is on the arc. Comparison made from the motion point back toward the origin is called the center-of-motion viewpoint and is used for relative physics.

Although real motion traverses the arc, internally the perception is the projection straight along the imaginary 'u_j'. However, this perception causes non-uniform rate.

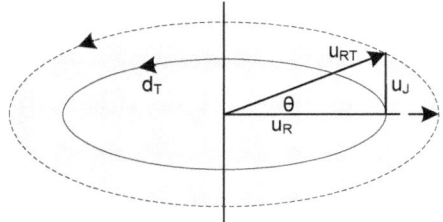

Figure 2 – Velocity Vectors

The forward velocity component 'u_j' is along the imaginary. The interface velocity 'u_{rt}' is the tip of the ray. When comparing arc motion to a point at the origin, the relative or line-of-sight velocity component 'u_r' along the axis can be calculated.

$$\frac{u_r}{u_{rt}} = \frac{\sqrt{u_{rt}^2 - u_j^2}}{u_{rt}}$$

Or,

$$cos\ \theta_{rt} = \frac{\sqrt{u_{rt}^2 - u_j^2}}{u_{rt}}$$

The trigonometric term is the power factor. The relativity factor 'rf' is the reciprocal of the cosine term.

When time is independently isolated, the factor becomes a ratio of distance vectors. This is a common term in dynamic energy conversion equations. It is a comparison of the ray and the line-of-sight along the reference axis.

$$rf = \frac{1}{cos\ \theta_{rt}}$$

$$rf = \frac{u_{rt}}{u_r}$$

So,

$$rf = \frac{b_{rs}}{b_r}$$

Now, compare the arc and the linear or imaginary projection.

$$u_t = u_{rt}\theta_{rt}$$

And,

$$u_j = u_{rt}\ sin\theta_{rt}$$

So,

$$\frac{u_t}{u_j} = \frac{\theta_{rt}}{sin\theta_{rt}}$$

The cosine term becomes the Lorentz transform when the radius velocity 'u_{rt}' near the interface is the speed of light.

Several observations are appropriate. Obviously, the forward linear velocity 'u_j' can never exceed the limiting ray velocity 'u_{rt}'. The actual arc velocity 'u_t' is greater than the linear velocity by 'θ_{rt} / sin θ_{rt}'. The arc velocity 'u_t' can be somewhat different from the ray velocity 'u_{rt}'.

The key component to the discussion is the elegance of relativity as well as the simplicity of Newtonian motion are both incorporated in the curve space and time divergence relationships of the unified field spheroidal coordinate system. The relativity concepts are as fundamental as the lever ray vector 'b_{rs}' relationship to the reference axis 'b_r'.

Summary _____

There are three unique distance types of rotational torque lever ray 'b_{rs}', motive wavelength 'd_t', and volume space 's'. Time has three manifestations of aeon or constant '$t=1$', chronos or cyclic motion 't_t', and kairos or seasonal reference 't_r'. These are implemented as independence '$t=1$', velocity '$1/t_t$', and acceleration '$1/t_t t_r$'.

Diffusion incorporates radiance as a change of curved area over time. Two velocities result from the rate of diffusion. These velocities move from divergent to parallel.

By judicious definition of the star operator, a simple field energy equation incorporates both lineal and rotational motion. Relative as well as Newtonian motion is contained in the relationships.

Annex _____

Consider the mass change due to the spheroidal coordinates. Use the known rest mass 'm_0' along the reference axis. With this definition, the motive mass, 'm_{rs}', is perceived moving along the arc of the ray on the surface. The relationship between rest mass and motive mass can be determined from the motive and lever velocities.

$$\frac{m_0}{m_{rs}} = \frac{u_r}{u_{rt}}$$

Then,

$$m_{rs} = rf\,(m_0)$$

$$= \frac{m_0}{\cos\theta_{rt}}$$

$$= \frac{m_0}{\sqrt{c^2 - u_j^2/c}}$$

The cosine term is the Lorentz factor. The motive mass is Einstein's explanation for mass expansion as it speeds toward light.

The spheroidal space with the divergent times provides a ready vehicle for resolving curved space-time analysis as well as fields and dynamic energy considerations.

6

STATIC ENERGY AT NO TIME

Thought
*Static energy
has no time.*
MOD

Abstract _____

Static energy is based on the three regents of matter for gravitational mass, electrical charge, and magnetic pole strength. Without time energy is static. Static energy is two regents of similar kind operating in a medium with a coefficient at a ray distance over a spherical area.

Introduction _____

The previous three chapters looked at the Unified Field Triad energy and the necessary mathematics of a spherical structure. The next three chapters develop the Unified Field System energy. System energy begins with static energy in this chapter, conversion energy in the next, then storage energy of a second order system.

Static Space_____

By definition, static energy is a field relationship. Static energy is the interaction between two regents of similar kind. Therefore, they are displaced or separated by a distance and the effect on the other is dispersed through the medium between them. Fluid dynamics are simply a field problem with a more defined medium.

For Newtonian physics, the action on the regents may be lumped in a relatively small area. Then, the energy can be determined at a node. Therefore, the field medium and the distributed space are lumped together to define a term such as capacitance, resistance, and inductance or spring, damper, and mass for electrical and mechanical systems respectively.

The volume is a sphere defined with the reference or origin regent at the center and the active regent on the surface. The volume can be represented as the average area of the surface of a Riemann sphere and the dot product of the space reference vector.

In a uniform circular space, the area and volume would have the following average values.

$$A_r = s_s \times s_t$$

$$A_r = 4\pi s_r^2$$

Where,

$$V_r = s_s \times s_t \cdot s_r$$

$$V_r = 4\pi s_r^2 \cdot s_r$$

$$V_r = \frac{4}{3}\pi s_r^3$$

It is critical to have volume for distributed regents and dispersed regents such as fluids. For discrete regents, the space reference vector in the numerator will correlate with the volume distance dot product in the denominator to yield units of area in the denominator.

$$\frac{s_r}{V_r} = \frac{1}{A_r}$$

Because of the duality between the electric charge and the magnetic pole strength in the unified field dynamic energy, in the system energy there can be either the electric analysis or the magnetic term, but not both. A correlation for converting from one to the other will be developed in a later section.

Similarly, because of the duality between the mass and diffusion in the Unified Field Triad energy, the system energy relationship can be expressed as one or the other, but not both.

Static Energy _____

Static energy is a comparison of the position effect of one regent of matter on another of similar kind, independent of time.

Law of static energy
The static energy between two regents is proportional to a product of the reference regent, the active regent, a field medium coefficient, a lever ray vector, and a space reference vector and is inversely proportional to the space volume they occupy.

$$E = \frac{b_{rs}k_n n_0 \, n_r \, s_r}{V_r} = \frac{b_{rs}k_n n_0 \, n_r \, s_r}{s_s s_t s_r}$$

The regents 'n' are gravitational mass 'm', electric charge 'q', and magnetic pole strength 'p'.

Equation of static energy, field model

$$E(k) = \frac{b_{xs}k_m m_0 \, m_x \, s_x}{V_x} + \frac{b_{ys}k_q q_0 \, q_y \, s_y}{V_y} + \frac{b_{zs}k_p p_0 \, p_z \, s_z}{V_z}$$

The equation of static energy necessarily includes all three regents. However, in most analysis, two are relatively insignificant and only one is necessary for calculations.

Energy is force acting through a distance between objects.

$$E = b_{rs} * F_r$$

So,

$$F_r = \frac{k_n n_0 \, n_r \, s_r}{V_r}$$

For a discrete regent in a uniform field, the force is the familiar Newton and Coulomb laws.

$$F_r = \frac{k_n n_0 n_r}{A_r} = \frac{k_n n_0 n_r}{4\pi s_r^2}$$

Static Medium Coefficients _____

The field medium coefficients are material properties for static energy conditions. The medium coefficient is based on a universal constant 'k_0' and a relative value 'k_r' for the specific medium conditions.

$$k_n = k_r k_0$$

The values of the universal constants are shown in the table.

Coefficient	k_0	Value	Std units	Alt units
k_m	$4\pi \times \gamma_0$	$4\pi \times 6.67384 \times 10^{-11}$	N-m^2/kg^2	(m^2/s)-m/kg-s
k_q	$1/\varepsilon_0$	$36\pi \times 10^9$	N-m^2/cb^2	wb-m/cb-s
k_p	$1/\mu_0$	$1/(4\pi \times 10^{-7})$	N-m^2/wb^2	cb-m/wb-s

The duality with diffusion 'D' can be seen in the mass alternate units of 'm^2/s'.

The relative values change based on the medium conditions.

The universal constants for permittivity 'ε' and permeability 'μ' are related by the speed of light.

$$c = c_r c_0$$

$$c_0 = \frac{1}{\sqrt{\mu_0 \varepsilon_0}}$$

And,

$$c_r = \frac{1}{\sqrt{\mu_r \varepsilon_r}}$$

Static Capacitance _____

System energy types depend on the operation of time. Static has no time 't=1', conversion has velocity with time '$1/t_t$', and storage accelerates with time '$1/t_t t_t$'.

Static energy is the system energy that is not dependent on time 't=1'. The energy is called potential energy in some perspectives. The energy may also be called capacitive energy. The force of attraction is centripetal. The capacitive component is called a spring 'K', capacitor 'C', or inductance 'L' for mass, charge, and magnetism respectively.

For convenience and consistency, the capacity of the system to store static energy will be called capacitance 'C'. The conversion to mechanical spring and magnetic inductance should be obvious and will be illustrated later.

Capacitance is defined by the properties of the medium and the distribution of the regents. For a node or lumped element, the numerator space reference 's_r' aligns with the volume dot product to result in an area in the denominator.

$$E = \frac{b_{rs} k_n n_0 \, n_r \, s_r}{V_r}$$

So,

$$\frac{1}{C_n} = \frac{b_{rs} k_n s_r}{V_r} = \frac{b_{rs} k_n}{A_r}$$

Then,

$$E = \frac{n_0 \, n_r}{C_n}$$

The static energy for lumped or node parameters depends on the mass, charge, and magnetism.

Equation of static energy, node model

$$E(k) = \frac{m_0\, m_x}{C_m} + \frac{q_0\, q_y}{C_q} + \frac{p_0\, p_z}{L_p}$$

There is a defined correspondence between the static energy and potential energy. They are related through the component value.

Static Capacitance for Mass _____

There are numerous ways the terms of the static energy can be grouped, dependent only on the definition of the desired parameter.

Equation of static energy, mass field model.

$$E(k) = \frac{b_{xs}k_m m_0\, m_x\, s_x}{V_x}$$

The static energy term for mass is used in at least three different forms. The first form is the standard capacitance relationship.

Equation of static energy, mass capacitance.

$$E(k) = \frac{m_0\, m_x}{C_m}$$

Where,

$$\frac{1}{C_m} = \frac{b_{xs}k_m s_x}{V_x} = \frac{b_{xs}k_m}{A_x}$$

The regent relationship for capacitance is determined by balancing the dynamic energy and static energy. Consequently, capacitance is the ratio of mass to the rate of diffusion.

$$\frac{m_x D}{t_t} = \frac{m_0\, m_x}{C_m}$$

So,

$$C_m = \frac{m}{D/_{t_t}}$$

The most common form realizes gravity using earth as the reference mass with its dimensions. The lever ray is the height above the earth surface. This is the familiar potential energy form.

Equation of static energy, mass gravity

$$E(k) = b_{xs}\, m_x\, g$$

Where,

$$g = \frac{k_m m_0 s_x}{V_x} = \frac{k_m m_0}{A_x}$$

The gravitational acceleration 'g' represents the intensity of the gravitational field of the earth or other reference mass.

Another common usage is the Newtonian form with a spring. In the interest of completeness, the following relationship is stated.

Equation of static energy, mass spring

$$E(k) = K\, b_{xs} s_x$$

Where,

$$K = \frac{k_m m_0\, m_x}{V_x}$$

The form uses distance as the regent, rather than mass. This is correlated in a later chapter where diffusion is developed as a component for static, conversion, and storage.

Static Capacitance for Charge_____

The second static energy term, with charge, is the basis for the electric circuit element of capacitance.

Equation of static energy, electric charge field model

$$E(k) = \frac{b_{ys}k_q q_0\, q_y\, s_y}{V_y}$$

As developed in the generic regent form, the static energy for an electric charge can be lumped into a node model.

Equation of static energy, electric charge

$$E(k) = \frac{q_0\, q_y}{C_q}$$

Where,

$$\frac{1}{C_q} = \frac{b_{ys}k_q s_y}{V_y} = \frac{b_{ys}k_q}{A_y}$$

The regent relationship for capacitance is determined by balancing the dynamic energy and static energy. Consequently, capacitance is the ratio of charge to magnetism rate, called voltage.

$$\frac{p_z q_y}{t_t} = \frac{q_0\, q_y}{C_q}$$

So,

$$C_q = \frac{q}{p/t_t}$$

Static Capacitance for Poles_____

The static energy equation depends on magnetism, as previously illustrated. The third static energy term for capacitance, with poles, is the basis for the magnetic circuit element inductance 'L'.

Equation of static energy, magnetic pole field model

$$E(k) = \frac{b_{zs}k_p p_0\, p_z\, s_z}{V_z}$$

As developed in the generic regent form, the static energy for an electric charge can be lumped into a node model.

Equation of static energy, magnetic pole

$$E(k) = \frac{p_0\, p_z}{L_p}$$

Where,

$$\frac{1}{L_p} = \frac{b_{zs}k_p s_z}{V_z} = \frac{b_{zs}k_p}{A_z}$$

The reciprocal of inductance of the magnetic circuit is conventionally called reluctance '$\mathcal{R}$'.

$$\mathcal{R} = \frac{1}{L_p}$$

The regent relationship for capacitance is determined by balancing the dynamic energy and static energy. Consequently, inductance is the ratio of pole magnetism to charge rate called current.

$$\frac{p_z q_y}{t_t} = \frac{p_0\, p_z}{L_p}$$

So,

$$L_p = \frac{p}{q/t_t}$$

Summary _____

Static energy is independent of time. The static energy equation is one triad of the Unified Field System. Each term represents one of the three regents of matter - mass 'm', charge 'q', and magnetism 'p'.

Equation of static energy, field model

$$E(k) = \frac{b_{xs}k_m m_0\, m_x\, s_x}{V_x} + \frac{b_{ys}k_q q_0\, q_y\, s_y}{V_y} + \frac{b_{zs}k_p p_0\, p_z\, s_z}{V_z}$$

$$\Leftarrow \Uparrow \Rightarrow$$

7

CONVERSION ENERGY IN TIME

Thought
Conversion energy changes form
but is neither created nor destroyed.
MOD

Abstract _____

Conversion is the second energy concept of the system energy. Conversion energy exists when energy changes form. Conversion energy involves one time change in the denominator. There are two unique conversion terms between electric - magnetic energy and mass - diffusion energy. Electric-magnetic conversion is based on static energy, while mass-diffusion storage comes from dynamic energy. Temperature and Boltzmann's constant comprise the third conversion term.

Conversion Energy _____

Energy is the common vehicle used for the conversion mechanism between dynamic forcing functions and system transitions. Energy can be described in terms of the fundamental matter, time, and space. As such, it can represent all that is known about physical systems.

The converted energy is derived from the time '1/t' change of energy. The energy may also be called resistive or viscous. The force of opposition is called resistance or friction. The lumped component is called a resistor 'R' or damper 'B' in electrical and mechanical systems respectively. Then the medium coefficient is resistivity 'r'.

Conversion energy is defined for each regent and diffusion. Conversion energy exists in the same complementary pairs found in the dynamic energy relationships. The duality between the pairs result is seen in the regent definition of resistances. Because of the duality in the dynamic equation, only a pair of conversion terms is unique. The third conversion term correlates to the constant term of the dynamic energy.

Electric-Magnetic Resistive Energy _

Conversion energy changes from one form of energy to another. A unique term is required to represent conversion from electric-magnetic energy.

> *Law of conversion energy, resistive field model*
> *The conversion energy between two regents is proportional to a product of the reference regent, the active regent, a field medium coefficient, the ray vector, and the space reference vector and is inversely proportional to the space volume they occupy and the time.*

Charge Resistive Energy _____

When the regent is charge, then the electric resistive energy is defined.

> *Equation of resistive energy, electric charge field model*

$$E(r) = \frac{b_{ys} r_q q_0 \, q_y \, s_y}{t_t \, s_s s_t s_y}$$

When the space reference vector is cancelled, the relationship is simplified.

$$E(r) = \frac{b_{ys} r_q q_0 \, q_y}{t_t \, s_s s_t}$$

The node or lumped energy equation combines the medium coefficient and the distance vectors.

Equation of resistive energy, electric charge node model

$$E(r) = R_q \frac{q_0 \, q_y}{t_t}$$

Where,

$$R_q = \frac{b_{ys} r_q}{s_s s_t}$$

Energy is force acting through a distance between objects.

$$E = b_{rs} * F_r$$

Where,

$$F_r = \frac{r_q \, q_0 \, q_y \, s_y}{t_t V_r} = \frac{r_q \, q_0 \, q_y}{t_t A_r}$$

The regent relationship for resistance is determined by balancing the dynamic energy and resistive energy. Consequently, resistance is the ratio of pole magnetism to electric charge.

$$\frac{p_z q_y}{t_t} = R_q \frac{q_0 \, q_y}{t_t}$$

So,

$$R_q = \frac{p}{q}$$

Pole Resistive Energy   _____

Pole resistive energy is the result of opposition to magnetic flux. It is not a unique term. The pole resistive energy can be developed directly from the regent form of charge resistive energy.

$$E(r) = R_q \frac{q_0 \, q_y}{t_t}$$

Where

$$R_q = \frac{p}{q}$$

So,

$$q = \frac{p}{R_q}$$

Substitution of the charge factor yields the resistive energy in terms of magnetic poles.

Equation of resistive energy, magnetic pole node model

$$E(r) = \frac{1}{R(q)} \frac{p_0 \, p_z}{t_t}$$

The pole resistance is the charge conductance. The duality of charge and poles is very apparent.

Diffusion Resistive Energy _____

Mass and diffusion resistive energy are friction in node applications and viscous in fluid applications. Diffusion is complementary to mass in energy.

Diffusion resistive energy is the result of opposition to area change over time. It is not a unique term. The diffusion resistive energy can be developed from the regent form of dynamic mass-diffusion energy. Diffusion resistive energy defines the action of a dashpot or damper 'B'.

Equation of resistive energy, diffusion node model

$$E(r) = B \frac{b_{rs} \, d_t}{t_t}$$

Alternatively, the motive distance over time is velocity.

$$E(r) = b_{rs} \, B \, u$$

Where,

$$F(r) = B \, u$$

The regent relationship for resistance is determined by balancing the dynamic energy and resistive energy. Consequently, damper resistance 'B' is the ratio of mass to motive time which is mass flow.

$$\frac{m_x D}{t_t} = B \frac{b_{rs} \times d_t}{t_t}$$

Or field model,

$$\frac{m_x}{t_t} \frac{b_{rs} d_t s_x}{t_r s_x} = B \frac{b_{rs} \times d_t}{t_t}$$

So,

$$B = \frac{m_x s_x}{t_r s_x}$$

Note that in fluid systems, the diffusion is related to kinematic viscosity. Furthermore, in some systems the mass is distributed over a volume to describe a density.

Mass Energy Variations _____

Consider the myriad variations and arrangements that are available for the dynamic energy mass diffusion term. This section is an illustration of the variations. A thorough development of these concepts is provided in other chapters.

$$E(d) = \frac{m_x}{t_t} \frac{b_{rs} d_t s_x}{t_r s_x}$$

$$= \frac{m_x D}{t_t}$$

Where,

$$D = \frac{b_{rs} d_t s_x}{t_r s_x}$$

For fluid systems, the dynamic energy mass term is expressed as pressure and volume.

$$E(d) = P\,V$$

Where,

$$P = \frac{m_x}{t_r t_t s_x}$$

As another example, the energy per volume is mass over space reference vector and divergent times. This is also the dynamic viscosity 'μ' rate.

$$\frac{E(d)}{V} = \frac{m_x}{t_r t_t s_x} = \frac{\mu}{t_t}$$

Where,

$$\mu = \frac{m_x}{t_r s_x}$$

The dynamic viscosity is the damper coefficient on the space reference.

$$B = \frac{m_x s_x}{t_r s_x} = \mu\, s_x$$

Kinematic viscosity 'v' is related to diffusion through the motion vectors.

$$v = \frac{b_{rs} d_t}{t_t}$$

And,

$$D = \frac{b_{rs} d_t s_x}{t_r s_x}$$

Then, energy is the mass flow and kinematic viscosity product.

$$E(d) = \frac{m_x D}{t_t} = \frac{\nu\, m_x s_x}{t_r s_x}$$

Energy is the product of the dynamic viscosity, the kinematic viscosity, and the space reference vector.

$$E(d) = \mu\, \nu\, s_x$$

Numerous other combinations and variations of terms are bound in the conversion of the dynamic energy mass term.

Natural Temperature _____

Other than light, temperature is one most notable manifestations of energy. The final term in the universal energy equation is the natural temperature 'T', Boltzmann's constant 'h_B', and the number of interactions 'N'. At the molecular level, N is the number of molecules.

$$E(T) = N\, h_B\, T$$

The natural temperature relationship is fundamental for an ideal system. Since no system is ideal, the variations are manifest in the other energy terms, most notable as an entropy relationship.

A Thing Called Entropy _____

Energy is commonly converted from one form to another. Electrical, mechanical, and chemical systems are mechanisms for conversions.

There is opposition or resistance 'R' to the conversion by the system giving up the energy 'E(r)'. The conversion of energy to another form results in work 'W' or accomplishment which is recovered. However, each conversion has a loss that is payback to the universe. The energy loss is converted to entropy 'S' and temperature 'T'.

$$E(loss) = S\,T$$

$$E(r) - S\,T = W$$

Entropy is an always increasing value and the loss results in a temperature increase.

In many systems, entropy is not measured. The loss may be expressed as a percentage of the converted energy 'E(r)'.

Efficiency 'η' is the ratio of the work or recovered energy to the converted or input energy. Efficiency has a decimal value between 0 and 1.

$$\eta = \frac{W}{E(r)}$$

The entropy relationship is not part of the fundamental unified equation. However, it is necessary to yield the work that is available to another system. Therefore, it will be shown parenthetical.

Summary _____

Conservation of Energy declares energy cannot be created or destroyed, but may change form. There is opposition to the change. This opposition creates the conversion energy.

Conversion energy is the sum of the electric-magnetic resistance, the mass-diffusion resistance and the natural temperature energy. Entropy is not a fundamental component of the unified theory, but is the energy loss at each conversion. So it is shown parenthetical.

$$E(c) = \frac{b_{ys}\,r_q\,q_0\,q_y\,s_y}{t_t\,s_s\,s_t\,s_y} + \frac{m_x\,s_x}{t_r\,s_x}\frac{b_{rs}\,d_t}{t_t} + N\,h_B\,T\,(-ST)$$

$$\Leftarrow \Uparrow \Rightarrow$$

8

STORAGE ENERGY AND SECOND ORDER SYSTEMS

Thought
*Second order is the triad
of physical systems.*
MOD

Abstract

Storage energy is not unique but rather is derived from another energy form that provides the impetus to storage. Electric-magnetic storage comes from static energy, while mass-diffusion storage comes from dynamic energy.

Second order systems contain three terms that are time related. The terms depend on the influence of time independence, time, and time again. These are the same form as the static, conversion, and storage energy values. Second order systems have a precise form. Therefore, a consistent solution is available.

System energy

System energy is the static, conversion, and storage elements due to mass, charge, and magnetism under the influence of time independence, time, and time again. Time independence is a static condition. Time causes opposition or resistance with motion, divergent time creates inertial or storage effect due to outside influence. System energy is the characteristic equation for the element organization of matter.

Energy states consist of *dynamic*, *static*, and *conversion*. The states are the components of the Unified field. System energy consists of three concepts of *static*, *conversion*, and *storage*.

Two of the energy states correspond to two of the system concepts. Therefore, it is reasonable to expect the storage concept to correlate in some form to the dynamic state. The dynamic state is the driving energy. Storage is the retained energy.

The storage energy for electric-magnetic and for mass-diffusion are illustrated. However, note these terms are derived from the dynamic energy for mass-diffusion and static energy for charge-poles. Therefore the terms are not unique.

Storage Charge from Static _____

The storage energy is derived from the second divergent time '$1/t_r t_t$' change on energy. The force of storage is centrifugal. The component is called an inductor, inertance, or mass inertia.

The storage energy is not a unique term. The storage energy can be described from the static node form of the complementary regent. The charge storage energy is the pole static energy. The charge storage energy is the result of the inertia of charge.

Equation of static energy, magnetic pole node model

$$E(k) = \frac{1}{L_p} \, p_0 \, p_z$$

The regent relationship for inductance 'L' is determined by balancing the dynamic energy and static energy. Consequently, inductance is the ratio of pole magnetism to charge rate called current.

$$\frac{p_z q_y}{t_t} = \frac{1}{L_p} \, p_0 \, p_z$$

Where

$$L_p = \frac{p}{q/t}$$

So

$$p = L_p \, \frac{q}{t}$$

Substituting the definition of poles into the pole static energy equation yields the charge storage energy equation.

Equation of storage energy, electric charge node model

$$E(l) = L_p \frac{q_0 \, q_y}{t_r \, t_t}$$

The duality between static poles and inductive charge has been illustrated. Similarly, the duality from the static charge to the inductive pole can be illustrated.

Equation of static energy, charge node model

$$E(k) = \frac{q_0 \, q_y}{C_q}$$

The regent relationship for inductance 'C' is the ratio of charge to poles over time called voltage.

$$C_q = \frac{q}{p/t_t}$$

Substituting in the static charge energy gives the storage pole energy.

Equation of storage energy, magnetic pole node model

$$E(l) = C_q \frac{p_0 \, p_z}{t_r \, t_t}$$

Storage energy does not involve any new terms.

Mass: Potential, Viscous, Kinetic ___

System energy has static, conversion, and storage concepts. The mass diffusion duality creates energy relationships based on the mass regent or the complimentary diffusion.

When mass is the basis for the static, conversion, and storage elements, the energy becomes potential, viscous, and kinetic respectively.

Potential energy is a representation of static mass.

$$E(k) = \frac{b_{xs} k_m m_0 \, m_x \, s_x}{V_x}$$

Potential energy (PE) is positional separation between two masses. Gravity uses a reference mass, such as earth for the comparison. The lever ray is the height above the reference earth surface.

$$PE = b_{xs} \, m_x \, g$$

Where,

$$g = \frac{k_m m_0 s_x}{V_x} = \frac{k_m m_0}{A_x}$$

Conversion energy is a representation of dynamic energy triad form. Several variations are noted in the previous chpter.

$$E(r) = \frac{m_x D}{t_t}$$

Storage energy is a representation of dynamic energy field form.

$$E(d) = \frac{m_x}{t_t} \frac{b_{rs} d_t s_x}{t_r s_x}$$

The peak and instantaneous kinetic energy results when the distance vectors are time related.

$$KE = m_x u_{rt} u_t$$

Diffusion: Spring, Damper, Mass____

System energy has static, conversion, and storage concepts. When diffusion is the basis for the static, conversion, and storage elements, the energy becomes spring, damper, and inertial respectively.

Spring potential energy is the spring constant 'K' on the dynamic motion variables. The spring constant is the rate of mass flow or mass over divergent times.

$$E(k) = \frac{m_x}{t_t} \frac{b_{rs} d_t s_x}{t_r s_x} = K \, b_{rs} d_t$$

Where,
$$K = \frac{m_x s_x}{t_r t_t s_x}$$

Damper conversion energy is the damper coefficient 'B' on the dynamic motion variables over time. The damper is a mass flow variation.

$$E(d) = \frac{m_x}{t_t} \frac{b_{rs} d_t s_x}{t_r s_x} = B \frac{b_{rs} d_t}{t_t}$$

Where,
$$B = \frac{m_x s_x}{t_r s_x}$$

Storage energy is the storage coefficient inertial mass 'm' on the dynamic motion variables over divergent times. The storage element is the mass.

$$E(l) = \frac{m_x}{t_t} \frac{b_{rs} d_t s_x}{t_r s_x} = m(l) \frac{b_{rs} d_t}{t_r t_t}$$

Where,

$$\frac{m_x s_x}{s_x} = m(l)$$

The dynamic energy is expressed in a variety of ways for the static, conversion, and storage elements of the system energy.

1. In the static energy, there is no time expressed, so the coefficient is mass and the space reference over two divergent times and the space reference.

2. In the conversion energy, there is one time expressed, so the coefficient is mass and the space reference over one time and the space reference.

3. In the storage energy, there is two times expressed, so the coefficient is simply mass and the space reference over the space reference.

For a second order system, there must be two similar variables. These are embedded in the mass-diffusion duality.

Second Order_____

Second order systems have a term with two time values. In the unified field, the system energy is second order with a static term, conversion term, and storage term.

In typical motion, the three terms would be distance with time independence, velocity which is distance over time, and acceleration which is distance over time again.

Transients are waveforms that exist for a short period of time. Waveforms are determined by the system elements. Since there are only three elements, the most complex network is second order. The characteristic solution for a system is the time varying equation that describes the exponential decay after a signal is applied.

Second Order Diffusion_____

A fundamental illustration of a second order system is a mass diffusion energy. The first term is the static energy, the second term is the conversion energy, and the third term is the storage energy derived from the dynamic energy.

$$E(t) = Kb_{rs} d_t + B \frac{b_{rs} d_t}{t_r} + m \frac{b_{rs} d_t}{t_r t_t}$$

Where

$K = static\ element$
$B = damper\ element$
$m = storage\ element$

As noted in an earlier chapter, when the components are aligned and time is short, the relationship can be reduced to a single distance 'b_{rs}' variable and the same time 't_r'.

Equation of second order system for distance

$$E(t) = Kb_{rs}^2 + B \frac{b_{rs}^2}{t_r} + m \frac{b_{rs}^2}{t_r^2}$$

The desired result is to know how the distance 'b_{rs}' changes as time varies from time=0 to the end of the analysis. That would be the solution to this second order equation.

Equation of second order system solution

$$b_{rs}(t) = F + (I - F)e^{-t/\tau} \cos(\omega t + \theta)$$

The terms have very important relationships defined by the second order system equation. Many of the values can be determined almost by inspection of the input conditions 'I' and the ending condition 'F'.

$F = final\ value\ (t = \infty)$
$I = initial\ value\ (t = 0)$

$t = time\ that\ varies\ from\ 0\ to\ \infty$
$\tau = time\ constant$
$\omega = natural\ frequency$
$\theta = phase\ shift\ angle$

'F' is the final value. If the energy varies with time, then the final value is the last value of the energy. In many cases, the final value goes to zero. If the energy has a constant value, then the final value is dependent on the source over the conversion 'B'.

$$F = \frac{E}{B}$$

I is the initial value. If there is no initial energy, then I is zero. If there is a source, then that is often the initial energy.

Time constant is the time it takes for the exponent of 'e' to be minus one, as the signal settles toward stability. The value of the time constant depends on whether the spring 'K' or the inertial mass 'm' term dominates.

$$\tau = \frac{m}{B}$$

Or,

$$\tau = BK$$

Frequency is inversely related to time. Angular frequency is one complete revolution of cycle of the frequency. Frequency of oscillation is independent of the damper.

$$\omega = \frac{1}{\sqrt{mK}}$$

A first order system has a damper and either a static or storage element. Therefore, there is no oscillation and the cos (0) is 1. However, there is still a cut-off frequency that is the inverse of the time constant.

Stability is reached when the varying value approaches the final value. Stability is 95% reached after 3 time constants and is 99% reached after 5 time constants.

The solution of the second order takes the form shown in the curve.

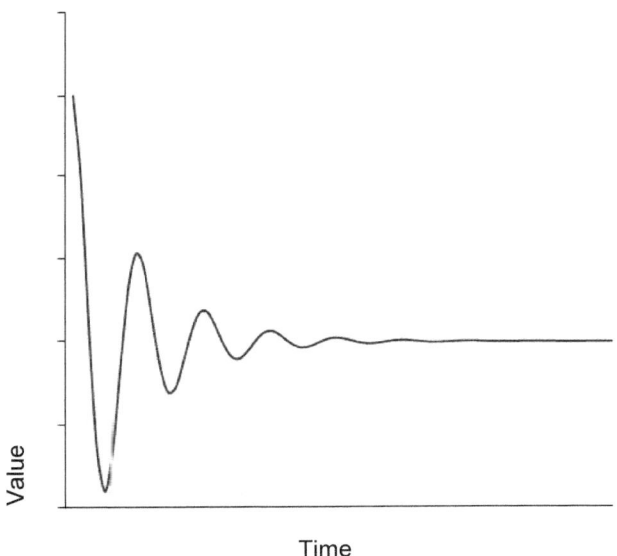

Plot of value of B as time varies from 0 to ∞

The solution is an oscillating exponential decay to a final value.

Second Order Electric Charge _____

The second order electric system equation is based on the charge. The first term is the static energy, the second term is the conversion energy, and the third term is the storage energy derived from the pole static energy.

$$E(q) = \frac{1}{C_q} q_0 \, q_y + R_q \frac{q_0 \, q_y}{t_t} + L_p \frac{q_0 \, q_y}{t_r \, t_t}$$

The electric circuit is obtained when the energy is equal to the electric-magnetic dynamic energy.

$$E(d) = \frac{p_z q_y}{t_t}$$

$$\frac{p_z q_y}{t_t} = \frac{1}{C_q} q_0 \, q_y + R_q \frac{q_0 \, q_y}{t_t} + L_p \frac{q_0 \, q_y}{t_r \, t_t}$$

If the equation is divided by the reference charge 'q', then the equation becomes the voltage relationship.

$$\frac{p_z}{t_t} = \frac{1}{C_q} q_y + R_q \frac{q_y}{t_t} + L_p \frac{q_y}{t_r \, t_t}$$

Where,

$$\frac{p_z}{t_t} = voltage$$

And,

$$\frac{q_y}{t_t} = current$$

Second Order Magnetic Pole _____

The second order magnetic system equation is based on the pole strength. The first term is the static energy, the second term is the conversion energy, and the third term is the storage energy derived from the charge static energy.

$$E(p) = \frac{1}{L_p} p_0 \, p_z + \frac{1}{R(q)} \frac{p_0 \, p_z}{t_t} + C_q \frac{p_0 \, p_z}{t_r \, t_t}$$

The magnetic circuit is obtained when the pole energy is equal to the electric-magnetic dynamic energy.

$$\frac{p_z q_y}{t_t} = \frac{1}{L_p} p_0 \, p_z + \frac{1}{R(q)} \frac{p_0 \, p_z}{t_t} + C_q \frac{p_0 \, p_z}{t_r \, t_t}$$

If the equation is divided by the reference pole strength 'p', then the equation becomes the current relationship.

Second Order Physics _____

Second order systems for fluid flow and chemical systems can be developed similar to the above concepts.

Second order is a common realization for all physical systems based on the triad principle. The relationship has been defined in economic as well as biologic systems. All these systems have the same response as that illustrated for the diffusion system.

Summary _____

Storage energy is the second order function of system energy. The terms are not unique. The electric-magnetic is derived from the static energy. The mass function is the dynamic energy mass term.

Second order systems have a static time independent term, conversion time dependent term, and storage time again term. The solution of the second order system has a cyclic exponent decay to a final value.

$$b_{rs}(t) = F + (I - F)e^{-t/\tau}\cos(\omega t + \theta)$$

$$\Leftarrow \Uparrow \Rightarrow$$

9

UNIVERSAL CONSTANTS

Thought
Universal constants
are natural fixed values.
MOD

Abstract _____

Universal constants are fixed values which exist in nature. The values aid in the calculation of specific types of energy. There are three constants associated with the Unified Field Triad energy. The Unified Field System energy includes the triad terms, plus three medium coefficients, and the temperature constant.

Introduction _____

The universal triad constants are imbedded in the Unified Field Triad energy. As would be expected, there are three and only three constants in the fundamental energy.

Pi 'π' is the ratio of the circumference to the diameter.

Speed of light 'c' is a constant circle constraint that forms a spheroidal boundary interface on velocity. So, it is the limiting velocity at the interface boundary. The speed of light binds cyclic time to motive distance, called wavelength.

Planck's constant 'h_P' is the energy value of a wave or quanta.

The complete suite of universal system constants arise from implementation of the unified field system equation. Only seven fundamental constants exist. These comprise the three triad constants, three medium constants, and temperature constant.

Permittivity 'ε' is the medium parameter for electric charge.

Permeability 'μ' is the medium parameter for magnetic poles.

Universal gravitation 'γ' is the medium parameter for mass.

Boltzmann's constant 'h_B' is the energy value of temperature.

Triad Constants_____

Three constants are noted in the Unified Field Triad energy system equation.

Pi 'π' is the ratio of the circumference to the diameter of a circle.

This was the first of the constants identified historically. The approximate value was known over 4000 years ago. By 1900 BC, the Babylonians and the Egyptians had identified a number. The value was known within 1% of the current accepted value. The approximate value is 3.14159. The value has been calculated to 5 trillion digits.

Pi is a non-repeating number. The value is estimated by calculating the length of the sides of an n-sided polygon that can fit inside a circle.

Speed of light 'c' is the limiting velocity at the interface boundary. As early as 400 BC, light was proposed to have a finite speed. The first recorded value was about 1676.

The value is about 299,792,458 meters per second. This is approximately 3×10^8 meters per second or 186,000 miles per hour.

Planck's constant 'h$_P$' is the energy value of a wave or quanta. Max Planck proposed the concept by 1900.

The value is h$_p$ = 6.626 069 57 x 10^{-34} J-s.

Medium Coefficients _____

Three constants describe the medium for mass, charge, and poles in the static condition. The constants are described for a vacuum, noted by the subscript '0'. The actual value required is the vacuum number multiplied by a relative number.

Universal gravitation 'γ' is the medium parameter for measurement between two masses. The constant was first mentioned by Sir Isaac Newton, but the value was not identified until 1798 by Henry Cavendish.

Permittivity 'ε' is the medium parameter for measuring between two electric charges.

Permeability 'μ' is the medium parameter for measuring magnetic pole strength.

The value of the coefficients in a vacuum are noted. Numerous different units are used. These are the SI values.

Coefficient	k$_0$	Value	Std units	Alt units
k$_m$	4π x γ$_0$	4π x 6.67384 × 10^{-11}	N-m^2/kg^2	(m^2/s)-m/kg-s
k$_q$	1/ε$_0$	36π x10^9	N-m^2/cb^2	wb-m/cb-s
k$_p$	1/μ$_0$	1/(4π x10^{-7})	N-m^2/wb^2	cb-m/wb-s

Temperature Constant _____

The seventh universal constant relates temperature to energy through the Boltzmann's constant. Although Boltzmann proposed the concept in 1877, Max Planck developed the first value about 1900.

The value is $h_B = 1.3806488 \times 10^{-23}$ Joule / Kelvin.

Other Constants _____

Other parameters have been proposed as universal constants. These indeed have great value in addressing certain problems. Some parameters, like Avogadro's number, are ubiquitous in chemistry, physics, or engineering. Nevertheless, only seven constants are necessary to define the Unified Field energy.

10

UNIFIED FIELD SUMMARY

Thought
*Unified field incorporates all
of matter, space, and time.*
MOD

Abstract _____

The Unified Field has two realizations. The Triad energy is a single equation for dynamic energy consisting of three terms. It encompasses all the measures of matter - mass 'm', charge 'q', and pole strength 'p'. The space time continuum is incorporated. The System energy includes three concepts of energy: dynamic, static, and conversion.

Triad Principle_____

Natural physical systems yield a principle for research models. The fundamental principle has been uniformly observed and ascertained for all physical relationships studied.

Triad Principle – Any item that can be uniquely identified can be further explained with three components.

Corollary – Two of the items will appear similar and the third will appear orthogonal.

Terms Recap _____

To this point all the terms have been introduced.

1. Energy consists of space, time, and matter.
2. Spheroidal space consists of three dimensions of tangential 's_t', standing 's_s', and radial reference, 's_r'.
3. Dynamic volume incorporates three unique space distance types of lever or ray 'b_{rs}', motive or wavelength 'd_t', and volume space 's_r'.
4. Similarly, time has three manifestations of constant '$t=1$', cyclic motion 't_t', and seasonal reference 't_r'.
5. Then, matter has three regents of mass 'm', electric charge 'q', and magnetic pole strength 'p'.
6. The life energy is the number of waves 'w' and Planck's constant of the universe 'h_p'.

Unified Field Triad Energy _____

The Unified Field must by definition incorporate all of matter, space, and time. The Unified Field Triad energy is the dynamic energy relationship.

Consistent with the Conservation of Energy and the triad principle, three energy terms are summed. At the boundary interface, the Unified Field provides the definition of light. The time space continuum is embedded in the diffusion.

Unified Field Triad Energy, node model

$$E(U) = \frac{m_x D}{t_t} + \frac{p_z q_y}{t_t} + \frac{h_p w}{t_t}$$

The Unified Field Triad energy, field model is the dynamic energy field model. The field model incorporates the dynamic volume gradient in each term.

Unified Field Triad Energy, field model

$$E(U) = \frac{m_x}{t_t}\frac{b_{rs}d_t s_r}{t_r s_r} + \frac{p_z q_y}{t_t}\frac{b_{rs}d_t s_r}{s_s s_t s_r} + \frac{h_p w}{t_t}\frac{b_{rs}d_t s_r}{s_s s_t s_r}$$

Unified Field System Energy _____

The Unified Field System energy incorporates the relationships to explain activity in a system. Unified Field System energy is the three dynamic energy terms (d), the three static energy terms (k), and the three conversion terms (c).

$$dynamic\ energy = static\ energy + conversion\ energy$$

The Unified Field System energy is one equation of three regent terms for three states of energy.

Unified Field System Energy, field model

$$\frac{m_x}{t_t}\frac{b_{rs}d_t s_r}{t_r s_r} + \frac{p_z q_y}{t_t}\frac{b_{rs}d_t s_r}{s_s s_t s_r} + \frac{h_p w}{t_t}\frac{b_{rs}d_t s_r}{s_s s_t s_r} =$$

$$\frac{b_{xs}k_m m_0\, m_x\, s_x}{s_s s_t s_x} + \frac{b_{ys}k_q q_0\, q_y\, s_y}{s_s s_t s_y} + \frac{b_{zs}k_p p_0\, p_z\, s_z}{s_s s_t s_z} +$$

$$\frac{b_{ys}r_q q_0\, q_y\, s_y}{t_t s_s s_t s_y} + \frac{m_x\, s_x}{t_t s_x}\frac{b_{xs}d_t}{t_r} + N\, h_B\, T\, (-S\, T)$$

As a unified field, these energy relationships are valid for any and every analytical perspective, whether cosmic or sub-atomic and Newtonian or Einsteinian. Units and other restrictive factors have been avoided to preclude limitation of perception.

The unified field is expressed in the most fundamental terms to allow arrangement for any system whether electrical, mechanical, or chemical.

Time Space Continuum _____

The dynamic volume is defined by the ray, motion, and space reference distance vectors.

$$V_r = b_{rs} \times d_t \cdot s_r$$

A field is simply energy that is distributed in a fluid, whether liquid or gas. Interestingly, a recurrent concept is the dynamic volume gradient 'dvg'.

$$dvg = \frac{b_{rs} \times d_t \cdot s_r}{s_r}$$

The notion of the dynamic volume gradient over reference time 't_r' is now defined as diffusion 'D'.

$$D = \frac{b_{rs} \times d_t \cdot s_r}{t_r \, s_r}$$

The space reference is very intentionally maintained in both the numerator and denominator. It is necessary in order to have volume in the numerator and it is necessary in the denominator to identify fluids and fields problems.

Curved Space_____

Space is a three-dimensional spheroid with measurements on the surface.

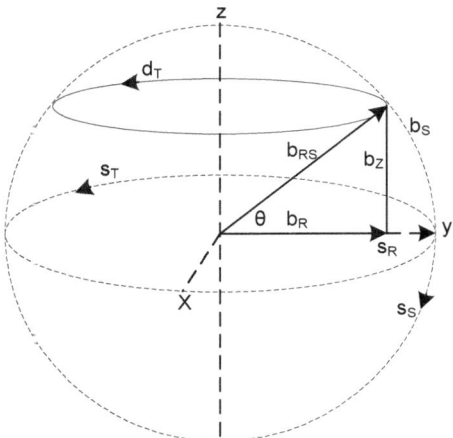

Figure 1 – Curved Space Vectors

The volume does not have to be uniform. The volume can be represented as the average area of the surface of a Riemann sphere and the dot product of the volume distance.

Energy, Force, & Rays _____

One concept of energy is the product of force through a ray distance. When that necessarily is expanded to incorporate spherical space vectors, then linear and rotational energy is defined by the star product.

$$
\begin{aligned}
E &= b_{rs} * F_r \\
&= b_r \cdot F_r + (b_s \times F_r)_t \\
&= b_r \cdot F_r + (\theta_{rs} b_j \times F_r)_t \\
&= b_r \cdot F_r + (T_t \theta_{rs})_t
\end{aligned}
$$

Each of the energy field terms has a ray distance vector 'b$_{rs}$' component. The remaining parameters of the term make up the force. The result is eight defined forces and the conversion temperature.

That's All _____

The desire for a unified field relationship is to tie together the concepts of matter, space, and time.

This elusive goal is realized with a very straightforward concept. As would be expected, the notion is expressed with three terms.

Annex A

TEMPERATURE AND THERMAL TRANSFER

Thought
*Temperature is one of the most
observable representations of energy.*
MOD

Abstract _____

Temperature is a natural consequence of energy. Temperature change results from energy conversion. Energy capacity involves the individual regents of matter without time. The transfer of energy, although not a fundamental component of the unified field, illustrates the movement of systems. Energy transport involves convection, conduction, and radiation.

Introduction _____

Other than light, temperature is one of the most observable representations of energy. This annex is an analysis of the temperature capacity and the thermal system transport.

Thermal Energy Capacity_____

The thermal capacity contains the inherent energy of the system without time influence 't=1'. The capacity coefficient is defined for a regent and material. This is a reversible relationship between temperature and energy. There is a correspondence between the thermal capacity terms and the dynamic energy because of the regents of matter.

Mass:

The thermal mass energy relationship is one of the foundations of the gas laws.

$$E = K_R \, m_x \, T$$

The first term for mass is the absorption heat. This incorporates the ideal gas law as well as material entropy.

The material coefficient 'K_R' is the specific heat. It is related through mass 'm' to the universal gas constant 'h_R' by the number of moles 'n'.

$$K_R \, m = n \, h_R$$

The universal gas constant 'h_R' is the product of Boltzmann's constant 'h_B' and Avogadro's number 'h_A'.

$$h_R = h_B \, h_A$$

So,

$$m = \frac{n \, h_B \, h_A}{K_R}$$

Electric-magnetic: The second thermal capacity term incorporates electric-magnetic processes.

Equation of thermal electric-magnetic energy

$$E = \frac{K_N \, p_z \, q_y \, T}{s_x \, s_y}$$

The material coefficient 'K_N' is related to the Nernst coefficient but it is charge dependent rather than current dependent. This is not a volumetric relationship.

The energy is rotational or circuital on the surface of the energy sphere. There is no traditional dot product type with force in line with a distance along an internal axis.

The energy is related to the Nernst-Ettingshausen properties. The energy is charge dependent while the Nernst type properties are current dependent. Nernst type is obtained when the energy change with time gives power or heat flow. This change causes a phase shift due to time and the associated wavelength unity factor 'd_t/s_t'.

When time changes, the relationship also includes the Righi-Leduc properties. The energy transfer is perpendicular to the temperature direction. It is independent of the separation between the hot and cold surfaces.

Other capacity:

One relationship for temperature capacity is included in the Unified Field System energy. This is not a new term, but is included for completeness.

The factor in the capacity energy relationship reflects the number of molecules, the wave temperature and the fundamental Boltzmann's constant, h_B.

Equation of thermal capacitance energy

$$E = N\, h_B\, T$$

This is directly related to the mass thermal term.

Thermal Energy _____

Power '$\mathcal{P}$' has been traditionally defined as the change of energy over time. This change causes a 90^0 phase shift due to time and the associated wavelength unity factor 'd_t/s_t'. Power is derived only when there is opposition to energy change and to flow rate. This results from conversion energy. In many applications, power is called heat transfer.

$$\mathcal{P} = \frac{E}{t}$$

Heat transfer is composed of conduction, convection, and radiation. All the terms transfer thermal energy across an area with its vector in the direction of heat flow. The area is the cross product of the lever arm 'b_{rs}' and the shear space 's_r'.

$$A_t = b_{rs} \times s_r$$

This area can take on any dimension. The most general is a circle or the surface of a spheroid.

The major difference between the forms is the relationship of the motive path or wavelength 'd_t'.

Thermal Transfer _____

Heat is the movement of thermal energy from a substance at a higher temperature to another substance at a lower temperature. Energy transfer is proportional to the product of a transport factor, surface area exposed, temperature change, and time.

$$E \propto k \; A T \, t$$

Transfer is frequently identified by power '$\mathscr{P}$' or heat rate. Power is obtained by dividing time from both sides.

$$\frac{E}{t} \propto k \, A \, T$$

A further refinement is power density, which is power over the area.

$$\frac{P}{A} = \frac{E}{A \, t} \propto k \, T$$

Heat is transported in three modes - convection, conduction, and radiation. Each mode has a different factor of transfer. The transport factor depends on the material property and the shape of the medium that is absorbing the heat. The heat is impinged on the perpendicular surface area, which is spheroidal.

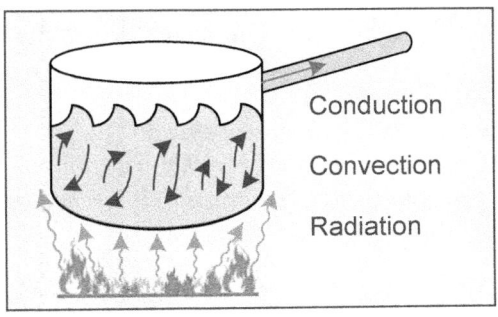

Transport_____

Convection is the transfer of heat by movement or circulation in fluids. Convection represents a phase change from solid to fluid. Hotter fluids rise while cooler fluids descend.

To a fluid moving by, it appears to be a shearing action. It is the transfer perpendicular to the direction of fluid motion. The area is parallel to the motion. The coefficient depends on the velocity, viscosity, material, and other properties.

The temperature change is the surface temperature minus the bulk temperature at a distance "far" from the surface. The factor 'k_V' must be derived or found experimentally for every system analyzed.

Equation of thermal convection energy
Convection energy is the product of transfer coefficient, area, temperature difference, and time.

$$E = k_V \, b_{rs} \, s_s \, \Delta T \, t$$

Conduction is the transfer of heat along a material by direct contact. Conduction depends on the mass of the material. Interestingly good electrical conductors tend to be good heat conductors. The temperature change is measured from one end of the material along a distance 'd' to the other end. The factor 'k_C' is the coefficient of conductivity.

Equation of thermal conductance energy
Conductance energy is the product of thermal conductivity property, area, time, and temperature difference over the motive distance.

$$E = k_C \, b_{rs} \, s_r \, \frac{\Delta T}{d} \, t$$

Radiation is the transfer of heat by waves. The transfer may be through a vacuum or any other medium. No mass is involved. The temperature difference is between the surface temperature and absolute zero.

Three variations exist for radiation – discrete, dynamic volume, and spectrum.

The factor 'h_B' is the Boltzmann's constant and the frequency is for a particular wave or wavenumber.

$$E = N\, h_B\, \Delta T$$

$$\mathcal{P} = N\, h_B\, f \Delta T$$

Radiation is the energy per volume. As such, it contains volume in its definition. The volume is the dot product of the cross sectional area and the motive distance or wavelength 'd_t'.

Wein's Law declares the product of the wavelength and temperature is constant.

Equation of thermal radiation energy
Radiation energy is the product of material coefficient, area, wavelength, temperature difference and time.

$$E = k_V\, b_{rs}\, s_r\, d_t\, \Delta T\, t$$

For a spectrum of radiation, the form becomes the Stefan-Boltzmann Law. Emissivity (ε) is a fraction of how well energy is radiated and sigma (σ) is the Stefan-Boltzmann constant. Since the emissivity is dependent on the material and structure, the radiation must be derived or found experimentally for every system analyzed.

$$\mathcal{P} = \epsilon\sigma A\, T^4$$

Although heat transport is well defined, the determination of the coefficients is less definitive. Therefore, models are subject to considerable experimental bias.

Summary _____

The thermal relationships are not fundamental to the Unified Field energy. Nevertheless, the temperature plays a critical part in energy and conversion.

Energy transfer involves convection, conduction, and radiation. The thermal capacity involves mass, charge-magnetism, and waves.

Annex B

THE CALCULUS, S'IL VOUS PLAÎT

> Thought
> *Calculus is an integral
> ingredient of uncertainty.*
> MOD

Abstract _____

The calculus has been assiduously avoided. The structure has been to provide a unified theory of physics with minimum mathematic constraints. This chapter is totally optional and is provided strictly to show the rigor that has been applied to the unified theory.

Introduction _____

The calculus has been assiduously avoided. Although calculus is very instructive in form, it unnecessarily creates conflict to understanding.

The approach has been to use the most fundamental terminology so the concepts are comprehendible by even introductory students of physics, who may not have the calculus mathematics background. Moreover, the equations are more discretely accurate without the calculus.

The equations and relationships provide the instantaneous or peak values. By definition of the structure of the Calculus, relationships are average values for a spectrum of data. For statistical analysis and uncertainty, that is an integral and necessary ingredient.

However, for a composite approach, the actual information at the instant of analysis is necessary. Therefore, averages and spectrums are less responsive.

Numerous techniques and procedures are typically used to reduce the complexity of the Calculus. These include transforms and numerical methods. The more common transforms are Laplace, phasor, and frequency domain.

With that caveat, the following sections are developed to validate the procedures used for the unified equation.

Spectrum and Sampling_____

The major difference between classical and atomic physics is in the perspective. Classical physics allows absolute measurement from a fixed reference. Because of the velocities, atomic physics tends to measure from a dynamic or relative reference. Both are valid compared to their perspective.

Where classical physics is primarily concerned with finding a number, atomic physics tend to look at a distribution over a range. Since the Calculus provides a differential perspective, it can be used to blend the two perspectives.

This is accomplished by equating the energy relationships that have discrete values. These include waves because of their cyclic nature, charges because of their defined size and rotation, and a thermal term for conversion.

The medium for transfer of energy can be environments such as waves, permeability, and permittivity. The medium 'w_n' is an integer multiple or harmonic 'w_r' on the fundamental value of the medium 'w_0'.

$$w_n = w_r \, w_0$$

The harmonic becomes the discrete sample number. The incremental energy 'E_r' is the harmonic on the fundamental energy 'E_0'. Energy between levels and photon energy between bundles, as well as other discrete types, is represented by the incremental energy.

$$E_r = w_r\, E_0$$

To determine a distribution, a comparison is made with the changing components of energy and number of samples as energy is converted from one type to another. The energy type that will be converted to 'E_x' times the difference between samples is equivalent to the number of samples 'Y' times the incremental energy converted.

$$E_x dY = Y\, E_r$$

Or,

$$\frac{dY}{Y} = \frac{E_r}{E_x}$$

Then,

$$Y_r = Y_0\, e^{-E_r/E_x}$$

The contribution of an individual sample 'Y_r' is a decay from the value of the fundamental sample 'Y_0'.

The energy contribution of a sample is the energy value on the sample value. Then energy contribution at any discrete sample point '$E_r Y_r$' is an exponential decay from the energy of the fundamental sample.

$$E_r\, Y_r = E_r Y_0\, e^{-E_r/E_x}$$

Boltzmann's distribution arises when the conversion is to thermal energy.

$$E_x = h_B T$$

The average energy over the entire spectrum is the summation of the energy at all the sample points divided by the summation of the sample points.

$$E_{avg} = \frac{\Sigma \, E_r \, Y_r}{\Sigma \, Y_r}$$

So,

$$E_{avg} = \frac{\Sigma \, E_r \, Y_0 \, e^{-E_r/E_x}}{\Sigma \, Y_r}$$

There is some obvious symmetry. By extensive manipulation, the resulting average energy is found.

$$E_{avg} = \frac{E_0}{\Sigma \, Y_r}$$

$$E_{avg} = \frac{E_0}{e^{-E_0/E_x} - 1}$$

Spectrum Radiation _____

Using classical mechanics, Lord Rayleigh determined the number of modes of oscillation per unit volume of space 'Y_d' in the wavelength range 'λ' to '$\lambda + d\lambda$'.

$$dY_\lambda = \frac{8\pi \, d\lambda}{\lambda^4}$$

The radiation or energy density 'D' in joules per cubic meter is the oscillation concentration on the average energy.

$$D = dY_\lambda \, E_{avg}$$

Then,

$$D = \frac{8\pi \, d\lambda}{\lambda^4} \frac{E_0}{e^{-E_0/E_x} - 1}$$

And,

$$D = \frac{8\pi\, E_0\, \lambda^{-4}}{e^{-E_0/E_x} - 1}\, d\lambda$$

This is a general relationship for radiation density. The fundamental energy can be acoustic, thermal, or photon, as well as other forms.

If the fundamental energy is Planck's frequency energy and the energy converted to is Boltzmann's thermal, then the relationship becomes Planck's emission law.

Integrating provides the Stefan-Boltzmann law of thermal radiation with the Stephan-Boltzmann constant 'σ'.

To convert to the energy distribution density to energy include the surface area and the emissivity factor '$\in$', which is a function of the surface characteristics.

$$E = \epsilon\, A_t\, \sigma\, T^4\, t_t$$

When the medium is a wave 'w', Then Planck's relationship develops for the fundamental energy.

$$E_0 = \frac{w_0\, h_P}{t_r}$$

So,

$$E_0 = h_P\, \frac{u}{\lambda}$$

Then,

$$E_0 = h_P\, \frac{c}{\lambda}\Big|_{interface}$$

The attenuation factor is the fundamental energy over the energy converted to.

$$z = \frac{E_0}{E_x}$$

$$z = \frac{c\, h_P}{h_B\, \lambda\, T}$$

$$z = \frac{h_D}{\lambda\, T}$$

The attenuation constant 'h_D' is speed of light on the ratio of Planck's to Boltzmann's constants.

$$h_D = \frac{c\, h_P}{h_B}$$

Radiation emittance 'P' in watts per cubic meter is the radiation or energy density 'D' over time.

Radiation intensity 'I' in watts per square meter is the velocity on the energy density 'D'. The velocity is the wavelength over time. When at the interface, the velocity is 'c'. For average values, divide by 4 to perform the Calculus compensation for the power of the wavelength.

REFERENCES

The source of some information is so accepted in the technical education, that the basis is not recalled. The fundamental relationships in various forms have been presented in papers for over 20 years. As a result, this book is built heavily on previous work.

The general references are noted. The following references clarified and helped formulate some concepts. They are not quoted in the traditional sense of a bibliography.

Papers resulting from the theory are also noted. The concepts were also used in the books written for university classes taught.

1. Angrist, Stanley W., *Direct Energy Conversion*, Allyn and Bacon, Boston, 1982.

2. Bergmann, Peter Gabriel, *Basic Theories of Physics*, Prentice-Hall, New York, 1949.

3. Chapman, Stephen J., *Electric Machinery Fundamentals*, McGraw-Hill, New York 1985.

4. Chirlian, Paul M., *Basic Network Theory*, McGraw-Hill, New York, 1969.

5. Durham, Marcus O., "A Composite Approach to Electrical Engineering", *Institute of Electrical and Electronic Engineers Region V*, 88CH25617-6/000-143, Colorado Springs, CO, March 1988, pp 143-147.

6. Durham, Marcus O., "A Universal Systems Model Incorporating Electrical, Magnetic, and Biological Relationships," *IEEE Transactions on Industry Applications*, Vol. 29, No. 2, March/April 1993, pp 436-446.

7. Durham, Marcus O., Robert A. Durham, and Karen D. Durham, "Applications Engineering Approach to Maxwell and Other

Mathematically Intense Problems", *Institute of Electrical and Electronics Engineers PCIC*, September 2002.

8. Durham, Marcus O., Robert A. Durham, and Karen D. Durham, "Applications Engineers Don't Do Hairy Math", *Proceedings of 35th Annual Frontiers in Power Conference*, OSU, Stillwater, OK, October 2002.

9. Durham, Marcus O., "Electromagnetics in One Equation Without Maxwell", *American Association for Advancement of Science - SWARM*, Tulsa, OK, April 2003.

10. Einstein, Albert, *Relativity - the Special and the General Theory*, authorised translation by Robert W. Lawson, Crown Publishers, New York, 1961.

11. Einstein, Albert, *The Meaning of Relativity*, Princeton University Press, Princeton, NJ, 1988.

12. Fazarine, Zvonko, "A Viewpoint on Calculus," *Hewlett-Packard Journal*, Palo Alto, CA, June, 1987. Previously presented to Mathematics Panel of American Association for the Advancement of Science.

13. Fink, Donald G. and John M. Carrol, Editor, *Standard Handbook for Electrical Engineers*, McGraw-Hill, New York, 1968.

14. Foster, Arthur R. and Robert L. Wright, Jr., *Basic Nuclear Engineering*, Allyn and Bacon, Boston, 1977

15. Holman, J. P., *Heat Transfer*, McGraw Hill, New York, 1990

16. Hoyt, William H., *Engineering Electromagnetics*, McGraw-Hill, New York, 1958.

17. Johnk, Carl T. A., *Engineering Electromagnetic Fields and Waves*, John Wiley, 1988.

18. Ogata, Katsuhiko, *System Dynamics*, Prentice-Hall, Englewood Cliffs, N.J. 1978.

19. Pilzer, Paul Zane, *Unlimited Wealth, The Theory of Economic Alchemy*, Crown Publishers, New York, 1990

20. Semat, Henry, and Robert Katz, *Physics*, Rinehart & Co., New York, 1958.

21. Wher, M. Russell, James A. Richards, and Thomas W. Adair III, *Physics of the Atom*, Addison-Wesley, Ready, MS, 1978.

22. Van Wylen, Gordon J., *Thermodynamics*, John Wiley, New York, 1964.

23. Yariv, Amnon, *Quantum Electronics*, John Wiley, New York, 1989.

ABOUT THE AUTHOR

Marcus O. Durham _____

Dr. Marcus O. Durham is a true polymath who brings very diverse experience to his writing and lectures. He is a scientist, philosopher, theologian, researcher, author, lecturer, entrepreneur, international consultant, applied psychologist, economist, energy scholar, engineer, forensic analyst, pilot, and professor emeritus.

He is a senior principal in an international scientific research lab, a principal in a forensic firm, and a principal in a natural resources company. He is a university professor emeritus of engineering, and was a seminary professor and dean.

He is a commercial pilot, is a ham radio Extra Class operator, has a commercial radiotelephone license, and is a licensed electrical contractor. He is a conservationist, who enjoys the family ranch and operating the heavy equipment.

Professional recognition includes Life Fellow-IEEE, Life Fellow-ACFEI, Diplomate-ABFET, Life Sr. Member-SPE, Certified Homeland Security, Certified Fire and Explosion Investigator, Certified Vehicle Fire Investigator, and Kaufmann Medal by IEEE.

Dr. Durham is acclaimed in *Who's Who of the Petroleum and Chemical Industry* and *Who's Who of American Teachers*. Honorary recognition includes Phi Kappa Phi, Tau Beta Pi, and Eta Kappa Nu.

He has published over 150 papers and articles and has authored 14 books on such diverse topics as science & engineering, economics & personal development, as well as philosophy & theology. He has traveled in over 22 countries and all 50 states.

Dr. Durham received the B.S. from Louisiana Tech University, the M.E. from The University of Tulsa, the Ph.D. in Engineering from Oklahoma State University, and the Ph.D. in Theology from Trinity

Southwest University. He has other studies with numerous educational and scholarly organizations.

The author has written and co-authored several books in the technical, philosophy, and applied psychology genres. Most are still available on-line or by order from bookstores.

- *Belief Tendencies, the Intersection of Science, Philosophy, & Theology,* Marcus O. Durham

- *Unified Field Theory in One Energy Equation,* Marcus O. Durham

- *Electrical Failure Analysis for Fire and Incident Investigation,* Marcus O. Durham, Robert A. Durham, Rosemary Durham, Jason Coffin

- *Electrical Engineering in a Nutshell,* Robert A. Durham, Marcus O. Durham

- *Electrical Systems – Fundamentals for Industry,* Marcus O. Durham, Robert A. Durham, Rosemary Durham, Jason Coffin

- *Leadership & Success in Relationships & Communication,* Marcus O. Durham, Robert A. Durham, Rosemary Durham

- *Leadership & Success in Organizations, Culture, & Ethics Culture,* Marcus O. Durham, Robert A. Durham, Rosemary Durham

- *Leadership & Success in Economics, Law, & Technology,* Marcus O. Durham, Robert A. Durham, and Rosemary Durham

- *An Intellectual's Argument About God,* Marcus O. and Rosemary Durham

- *Systems Design and the 8051,* Third Edition, Marcus O. Durham

- *Who Is This God?* Marcus O. and Rosemary Durham

- *Micro-Controllers in Systems Design,* Marcus O. Durham

- *Electrical Engineering Circuit Concepts,* Marcus O. Durham, Robert A. Durham

- *Electric Machines & Power,* Marcus O. Durham, Robert A. Durham

$$\Leftarrow \Uparrow \Rightarrow$$